AF521384

Foodgrain Market Emerging Issues

Foodgrain Market Emerging Issues

Vikram Sharma

RANDOM

RANDOM PUBLICATIONS

NEW DELHI (INDIA)

Foodgrain Market Emerging Issues

ISBN 978-93-5111-487-1
© Reserved

All Rights Reserved. No Part of this book may be reproduced in any manner without written permission.

Published in 2015 in India by

RANDOM PUBLICATIONS

4376-A/4B, Gali Murari Lal, Ansari Road
New Delhi-110 002
Phone : +9111-43580356, 011-23289044, 011-43142548
e-mail: sales@randompublications.com,
info@randompublications.com, randomexports@gmail.com

Reprinted, 2018

Type Setting by : Friends Media, Delhi-110089
Printed at : Replika Press Pvt. Ltd.

Preface

In spite of improved supply prospects and weakening demand, foodgrain market conditions remain fairly tight, which is the major factor underpinning prices. Production forecasts for nearly all key food crops have risen steadily. For cereals, larger anticipated inventories reflect not only improved production prospects but also expectations of a slowing demand growth because of the unfavourable macroeconomic environment. In spite of these developments, however, international prices of foodgrains continue to be high. Strong underlying demand in certain countries, where economic growth is robust, is price supportive. Aside from being high, most prices are also extremely volatile, moving in tandem with unstable financial and equity markets. Fluctuations in exchange rates and uncertainties in energy markets are also contributing to sharp price swings in agricultural markets.

Given all these uncertainties, it is difficult to predict how markets will evolve in the near-term. While there is some room for optimism that, for most commodities, prices could remain below their recent highs, the general picture still points to firm markets well into coming years. For most food commodities, future year's production will have to increase in order to meet the expected demand, albeit moderately. However, if this demand were to rise faster than currently envisaged, which is a possibility even assuming a slow economic recovery, then a more significant production expansion will be required. Input costs, from fertilizers to energy, remain high, interest rates have climbed in many emerging economies, all of which could dampen production and, hence, draw down stocks and boost prices further.

Reducing market uncertainty may not be among the fastest remedies for lowering the number of hungry. Yet, letting international markets continue in their present state, volatile and unpredictable, will only aggravate an already grim outlook for world food security. This is the reason why world leaders have been dwelling at length on the issue of price volatility. Such discussions gained momentum in recent months as attention turned towards finding ways to improve the accuracy of supply and demand forecasts for major food crops as an important first step in promoting stable and transparent food markets.

Foodgrain markets have changed rapidly in recent years, and while standard treatments of price analysis are still relevant, there is a critical need to train students to work in the new world of agricultural markets. This book incorporates new information and devotes more time and space to the issues of foodgrain market likely to be faced by applied economists. Providing students with the necessary background to understand the new environment in which they will be expected to operate, this book is indispensable reading for both upper-level undergraduates and graduate students.

Author

Contents

1

Agricultural and Food Marketing Systems

As individuals within a society become more specialised in their economic activities, they come to rely upon others to supply at least some of the products and services which they need. Thus begins a process of exchange between buyers and sellers. For a while buyers and sellers remain in immediate contact and each party is able to determine what the other needs and values and, therefore, will be willing to exchange. As the economy develops the number and types of exchanges expand, there is a concomitant need for increasingly specialised marketing services such as physical distribution, storage, grading, market information gathering and so. The number of participants also increases with many of the specialised services being provided by intermediaries between the seller and ultimate buyer. Few buyers and sellers are in direct contact with one another and communication between them is channelled through a complex marketing system.

Importance of Agricultural and Food Marketing

In many countries, and virtually every less developed country (LDC), agriculture is the biggest single industry. Agriculture typically employs over fifty percent of the labour force in LDCs with industry and commerce dependent upon it as a source of raw materials and as a market for manufactured goods. Hence many argue that the development of agriculture and the marketing systems which impinge upon it are at the heart of the economic growth process in LDCs. Moreover as Kriesberg points out; in

LDCs the consumer frequently spends in excess of fifty percent of the household's income on basic foodstuffs - much of which is inadequate both in quality and nutritional content. By contrast Americans spend approximately twelve percent of their total disposable income on food. In Western Europe the figure ranges from about sixteen to nineteen percent of disposable income. Furthermore, whereas in developed countries the poor are relatively few in number, and therefore it is economically possible to establish special food distribution programmes to meet their needs, the scale of poverty in most LDCs is such that the commercial marketing system must be relied upon to perform the task of food distribution to poor and not-so-poor alike. This being so, it is imperative that the marketing system performs efficiently.

Economic development itself provides the impulse towards more sophisticated and more efficient marketing systems. Dixie suggests that as countries experience economic growth, their rate of urbanisation tends to increase substantially. Whereas the rate of population growth, in developing countries, averages around three percent per annum, their cities and towns are increasing their populations at about four percent per annum. In essence, this means that the number of people, in urban areas, needing to be fed by rural people, will double within sixteen years. This has clear implications for agricultural production and the marketing systems that direct that production and distribute the output to the points of its consumption. Subsistence farming is likely to diminish in importance as farmers respond to the increased opportunities that development and urbanisation create; farms are likely to decrease in number whilst increasing in size; and agriculture will probably become less labour intensive and more capital intensive.

Dixie also highlights the potential contribution of agricultural and food marketing, towards attempts to improve rural incomes in developing countries. The inequality of incomes between the rural and urban areas draws people away from agricultural production and places great stress upon the infrastructure and social services of a country's towns and cities. Nowhere was this more dramatically demonstrated than in Nigeria when petroleum oil was discovered and then exploited in the 1970s. A large number of jobs were created in the urban areas and people abandoned agricultural production in large numbers. Nigeria became a net importer of many agricultural products of which it had formerly been a net exporter. For as long as the world price for petroleum remained high the economy thrived and could well afford the food import bill. However, as soon as the world price for oil fell, the food import bill became a serious burden. Nigeria would only have avoided this scenario if it had been able to motivate people to continue in agriculture and this would only have been possible if the disparity between urban and rural incomes had been reduced. Rurally based enterprises, including small-holdings, can greatly improve their earning potential by adopting a market orientation. They can be encouraged to add value to commodities by adding to their utility. Value added products normally carry a higher margin than raw commodities.

Another development which has in recent times increased interest in marketing practies is the trend, in many developing countries, towards market liberalisation as part of economic structural adjustment programmes (ESAPs). The view that direct and indirect government participation in production and distribution had brought about structural distortions in economies has become widely accepted. Measures intended to correct these distortions include a return to market prices for all products and resources, the encouragement of a competitive private sector and the commercialisation, and sometimes privatisation, of all or some of the functions of marketing parastatals. All of this requires a better understanding of marketing practices and processes within the country implementing ESAPs, in general, and within the agricultural marketing parastatals affected, in particular.

So far this discussion has been set in the context of commercial marketing but social marketing should also be acknowledged. Social marketing identifies human needs in non-competitive economies and/or sectors of society and defines the means of delivering products and services to meet these needs. The marketing mix of social marketing strategies is evaluated using quite different criteria from those employed in assessing purely commercial marketing strategies. Criteria such as the percentage of the target population reached with the technology, products, processes or services, quantities produced and distributed and uptake of the product, service or technology are more often employed. Benefits are measured in terms of development goals, such as improved nutritional status or increased rural incomes. The use of economic criteria is usually limited to the latter and to selecting the least-cost strategy to achieve a quantitative goal. However, the criteria used to evaluate commercial marketing strategies should not automatically be eliminated, because these improve the efficiency of some aspects of social marketing strategy without preventing the attainment of social objectives.

Marketing Concept and Marketing Systems

Marketing is not simply an extension of the production process but its only purpose as Adam Smith emphasised when, in his text *The Wealth of Nations* (1776), he said that:

> "Consumption is the sole end purpose of all production: and the interest of the producer ought to be attended to only so far as it may be necessary for promoting that of the consumer."

Dixie relates what he describes as a definition of marketing which is:

> "The series of services involved in moving a product (or commodity) from the point of production to the point of consumption."

This is a definition which many organisations, and governments, would recognise as describing their own activities in commodity marketing. Indeed in many developing countries it aptly describes, or in some cases, did in the past describe, the functions carried out by marketing parastatals with respect to staple foods. However, as Dixie himself points out, the definition omits two key elements of any definition of marketing production to effuse the marketing concept, i.e. a customer orientation and inbuilt sustainability. Gaedeke and Tootelian offer an alternative definition which overcomes the problems caused by these two omissions:

> "... a management orientation focusing all the activities of the organization on satisfying customer needs and wants, thereby helping achieve the organization's long-range objectives."

This definition promotes a customer orientation and since the organisation's long-term objectives will include it's own continued existence it takes account of the need for sustainability. Moreover, this definition of the marketing concept does not preclude non-profit making organisations. Marketing is just as relevant to development projects, aid agencies, extension service organisations, and the like, as it is to commercial enterprises. Thus the marketing concept is that an organisation achieves its goals through the provision of customer satisfaction. Put another way, marketing is the integrative force that matches production to customer needs and satisfaction.

Marketing is not an activity to which an organisation turns its attention at the end of the production phase of operations. Rather marketing needs to be directing production in accordance with clear signals from the marketplace as what is needed by customers.

The marketing concept must be adopted throughout not only the entire organisation, but the entire marketing system. A system is a complex of interrelated component parts or sub-systems which have a defined common goal. Thus, an agricultural and food marketing system comprises all of the functions, and agencies who perform those activities, that are necessary in order to profitably exploit opportunities in the marketplace. Each of the components, or sub-systems, are independent of one another but a change in any one of them impacts on the others as well as upon the system as a whole.

There is a danger that the marketing concept will be adopted by some parts of the system but not others. Thus, for example, a food manufacturer may be trying hard to implement the marketing concept and offer products that meet the precise needs of a target market. If, however, the manufacturer has to rely upon a farming community that is still very much production oriented, for raw material supplies, then the overall marketing objectives may be frustrated. In the same way, if only some functions are performed according to the marketing concept then the system as a whole may not achieve a market orientation.

For instance, the marketing department may set out to serve the market for a high quality fruits and vegetables, for which it can obtain premium prices, but if transportation is performed using the same open-topped bulk carrying wagons used to ship grain and other aggregates then it is unlikely that the enterprise will deliver the product in the right condition for the target market.

The lesson from all of this is quite simple. If the decision has been made to adopt the marketing concept, then consideration has to be given to the implications for each of the participants and the functions performed within the marketing system. Where one or more elements of the system are found to be other than market orientated, then either a change towards the marketing philosophy has to be introduced in those elements or a change in the configuration of the marketing system has to be implemented.

Marketing Sub-systems

Rosson conceives of agricultural and food marketing systems as consisting of 4 main sub-systems; production, distribution, consumption and regulatory. The key players in the chain of activities that connect food and agriculture are the farmer, intermediaries, the food processors, and the consumer. In practice they each see the agricultural/food marketing system from a perspective of self-interest and these interests are sometimes in conflict. Illustrative examples of some of the conflicts which typically arise are given in table 1.

Table 1. Conflict of interest in agricultural/food marketing systems

Key Players	*Interests*
Farmers	Maximum price, unlimited quantities
Manufacturers	Low purchase price, high quality
Traders and retailers	Low purchase price, high quality
Consumers	Low purchase price, high quality

The farmer's interest is focused on getting the best return from his produce, which usually equates to maximum price for unlimited quantities. Manufacturers want least cost, best quality produce from the farmer so that he can sell it at competitive, but profitable, prices. Traders and retailers want high quality and reliable supplies from the manufacturer or farmer, at the most competitive prices. Consumers are interested in obtaining high quality products at low prices. Clearly, there are conflicting interests here.

In an ideal world there should be some form of strategic partnership between these key players. It is obvious that, in the long run, any one of the four groups would find it difficult to survive if the others do not. However, in real life, attitudes are not those of the ideal world or of the longer term. It is focused more on the shorter term and in preserving the interests of each group. Only by allowing each group to take care of its interests, can a balanced longer term relationship evolve. This must be borne in mind when considering what the food industry expects from agriculture.

Marketing Functions

A little earlier it was said that a marketing system has two distinct dimensions. One of those dimensions is the institutions, organisations and enterprises which participate in a market and the second is the functions that those participants perform. Kohls and Uhl have classified the functions involved in agricultural and food marketing processes as under three sets of functions of a marketing system.

Table 2. Functions involved in agricultural and food marketing processes

A. Exchange Functions	1. Buying 2. Selling
B. Physical Functions	3. Storage 4. Transportation 5. Processing
C. Facilitating Functions	6. Standardisation 7. Financing 8. Risk Bearing 9. Market Intelligence

Each of these functions add value to the product and they require inputs, so they incur costs. As long as the value added to the product is positive, most firms or entrepreneurs will find it profitable to compete to supply the service.

Exchange Functions

Buying

The marketing concept holds that the needs of the customer are of paramount importance. A producer can be said to have adopted a market orientation when production is purposely planned to meet specific demands or market opportunities. Thus a contract farmer who wishes to meet the needs of a food processor manufacturing sorghum-based malted drinks will only purchase improved sorghum seed. He/she will avoid any inputs likely to adversely affect the storage and/or processing properties of the sorghum and will continually seek new and better inputs which will add further value to his/her product in the eyes of the customer. In making his/her buying decisions his underlaying consideration will be the effect upon the attractiveness of his/her output to the markets he/she is seeking to serve.

The buyer's motive is the opportunity to maintain or even increase profits and not necessarily to provide, for example, the best quality. Improving quality inevitably increases the associated costs. In some cases the market is insensitive to improvements in quality, beyond some threshold level, does not earn a premium price. Under such circumstances, the grower who perseveres and produces a 'better product', is not market oriented since he/she is ignoring the real needs of the consumer. The most successful agribusiness is the one which yields the largest difference between prices obtained and costs incurred.

Selling

Of the nine functions listed, this is probably the one which people find least difficulty in associating with marketing. Indeed to many the terms marketing and selling are synonymous. Kotler suggests that:

> "Most firms practice the selling concept when they have over capacity. Their immediate aim is to sell what they can make rather than to make what they can sell."

There is no denying that 'high pressure selling' is practiced, where the interests of the consumer are far from foremost in the mind of the seller. This is not marketing. Enterprises adopt the marketing philosophy as a result of becoming aware that their own long term objectives can only be realised by consistently providing customer satisfaction. Whereas selling might create a consumer, marketing is about creating a customer. The difference is that marketing is about establishing and maintaining long term relationships with customers.

Selling is part of marketing in the same way that promotion, advertising and merchandising are components, or sub-components of the marketing mix. These all directed towards persuasion and are collectively known as marketing communications; one of the four elements of the marketing mix.

Physical Functions

Storage

An inherent characteristic of agricultural production is that it is seasonal whilst demand is generally continous throughout the year. Hence the need for storage to allow a smooth, and as far as possible, uninterrupted flow of product into the market. Because he is dealing with a biological product the grower does not enjoy the same flexibility as his manufacturing counterpart in being able to adjust the timing of supply to match demand. It would be an exaggeration to suggest that a manufacturer can turn production on and off to meet demand - they too have their constraints- but they have more alternatives than does the agricultural producer. A manufacturer can, for example, work overtime, sub-contract work, and over a longer time horizon, the manufacturer can increase or decrease productive capacity to match the strength of demand.

In agriculture, and especially in LDCs, supply often exceeds demand in the immediate post-harvest period. The glut reduces producer prices and wastage rates can be extremely high. For much of the reminder of the period before the next harvest, the product can be in short supply with traders and consumers having to pay premium prices to secure whatever scarce supplies are to be had. The storage function is one of balancing supply and demand.

Both growers and consumers gain from a marketing system that can make produce available when it is needed. A farmer, merchant, co-operative, marketing board or retailer who stores a product provides a service. That service costs money and there are risks in the form of wastage and slumps in market demand, prices, so the provider of storage is entitled to a reward in the form of profit.

Transportation

The transport function is chiefly one of making the product available where it is needed, without adding unreasonably to the overall cost of the produce. Adequate performance of this function requires consideration of alternative routes and types of transportation, with a view to achieving timeliness, maintaining produce quality and minimising shipping costs.

Effective transport management is critical to efficient marketing. Whether operating a single vehicle or a fleet of vehicles, transportation has to be carefully managed, including cost monitoring - operations on different road types, fuel and lubrication consumption and scheduled and remedial maintenance and repair. Skillful management of all aspects of vehicle operations can also make a substantial contribution to efficient marketing especially with respect to optimum routing, scheduling and loading and off-loading; maximisation of shift hours available, maintaining the vehicle fleet at an optimum size, taking account of time constraints on delivery, and collection times and judicious management of vehicle replacement and depreciation. Transport managers also have to weigh the advantages and disadvantages of owning, hiring or leasing transport.

Processing

Most agricultural produce is not in a form suitable for direct delivery to the consumer when it is first harvested. Rather it needs to be changed in some way before it can be used. Kohls and Uhl observe that:

> "The processing function is sometimes not included in a list of marketing functions because it is essentially a form changing activity."

However, it is for this very reason that processing ought to be included as a marketing function. The form changing activity is one of that adds value to the product. Changing green coffee beans into roasted beans, cassava into gari or livestock feed, full fruit bunches into palm oil or sugarcane into gur increases the value of the product because the converted product has greater utility to the buyer. How the form of produce is to be changed and the method to be used in bringing about such changes are marketing decisions. For example, some years ago when Ethiopia was looking to expand its tea business, a prototype manufacturing plant was established. The plant was capable of curing the tea and packing it in individual tea bags. At that point, tests were undertaken in which the product was compared with others already on the market. The results were encouraging. However, in the course of the marketing research, it was also discovered that ninety percent of the black tea consumed is blended and not the pure variety placed in tea bags by the Ethiopians. By going past the point of changing green leaf into high quality black tea, the Ethiopians were entering a nice market which is not what they intended at all. Timely marketing research would have directed Ethiopia to stop the form changing activity short of bagging since, at that time, Ethiopia did not have the acreage of tea, nor the resources, to develop a tea blending facility of its own. In the same way, a producer of fresh fruits may have pulping and/or canning facilities but if potential buyers want the flexibility of using the fruits in a variety of ways, then these stages of processing serve to reduce utility and value, rather than increasing them.

Of course, processing is not the only way of adding value to a product. Storing products until such times as they are needed adds utility and therefore adds value. Similarly, transporting commodities to purchasing points convenient to the consumer adds value. In short, any action which increases the utility of the good or service to prospective buyers also adds value to that product or service.

Facilitating Functions

The facilitating functions include product standardisation, financing, risk bearing and market intelligence. Facilitating functions are those activities which enable the exchange process to take place. Marketing, in simple terms, is the act of supplying products to someone in exchange for something perceived to be of equal or greater value, (usually, but not always, a given sum of money). Facilitating functions are not a direct part of either the exchange of title or the physical movement of produce.

Standardisation

Standardisation is concerned with the establishment and maintenance of uniform measurements of produce quality and/or quantity. This function simplifies buying and selling as well as reducing marketing costs by enabling buyers to specify precisely what they want and suppliers to communicate what they are able and willing to supply with respect to both quantity and quality of product. In the absence of standard weights and measures trade either becomes more expensive to conduct or impossible altogether. In Nepal such was the diversity of weights and measures used with respect to grain within the country, that it was easier for some districts to conduct trade with neighbouring states in India than it was to do business with other districts within Nepal. Among the most notable advantages of uniform standards, are:

- price quotations are more meaningful
- the sale of commodities by sample or description becomes possible
- small lots of commodities, produced by a large number of small producers, can be assembled into economic loads if these supplies are similar in grade or quality
- faced with a range of graded produce the buyer is able to choose the quality of product he/she is able and willing to purchase.

Quality differences in agricultural products arises for several reasons. Quality differences may be due to production methods and/or because of the quality of inputs used. Technological innovation can also give rise to quality differences. In addition, a buyer's assessment of a product's quality is often an expression of personal preference. Thus, for example, in some markets a small banana is judged to be in some sense 'better' than a large banana; white sugar is considered 'superior' to yellow sugar; long stemmed carnations are of 'higher quality' than short stemmed carnations; and white maize is 'easier to digest' than yellow maize. It matters not whether the criteria used in making such assessments are objective or subjective since they have the same effect in the marketplace. What does matter in marketing is to understand how the buyer assesses 'quality'.

Financing

In almost any production system there are inevitable lags between investing in the necessary raw materials (e.g. machinery, seeds, fertilizers, packaging, flavourings, stocks etc.) and receiving the payment for the sale of produce. During these lag periods some individual or institution must finance the investment. The question of where the funding of the investment is to come from, at all points between production and consumption, is one that marketing must address. Consider the problem of a food manufacturer who wishes to launch a range of chilled products in a developing country where few retail outlets have the necessary refrigeration equipment. This is a marketing problem. It might be solved by the food manufacturer buying refrigerators and leading these to retailers (or arriving a hire-purchase arrangement with retailers).

A common marketing problem, in developing countries, is the low level of incomes leading to low levels of effective demand for many products. The challenge to marketing is to somehow channel what income is available into effective demand. In the case of agricultural equipment marketing this might involve offering hire-purchase schemes where the prospective buyer makes payment in regular installments. During this time he/she is deemed to have hired the machine. If payments are not forthcoming, the machine can be recovered since its ownership remains with the seller up until the final payment is made, at which point the farmer is considered to have purchased the machine.

Alternatively, the seller might set up leasing, rather than purchasing schemes where again the farmer is making regular payments but never takes title to the machine. Where a food item is being marketed, to a low income market, the seller can consider reducing the until price of the product by making the pack or lot size smaller.

Another tactic is to make the product more affordable by using cheaper ingredients and/or packaging. Instant coffee can be sold at lower prices by substituting some of the coffee with chicory; the price of meat products is reduced by increasing the percentage of cereals in these products and including less meat and/or making use of less expensive parts of the animal such as entrails, offal, feet and head.

Marketing is also concerned with the financing of the enterprise itself. Here again some creative solutions can be developed. Where internal financing is insufficient for the purposes in view, an enterprise in a developing country can look to several alternatives including:

- development banks
- commercial banks
- shares issues
- credit co-operatives and/or credit unions

Where these sources of finance are considered inappropriate, or are simply not available to a particular enterprise, a strategic alliance in the form of a joint venture could be the answer. These are partnerships formed to exploit market opportunities more effectively and/or efficiently than either party can on its own. An enterprise, in a developing country, may engage in a joint venture with either an indigenous partner and/or with a foreign partner. The agreement between parties to a joint venture normally specifies their respective contributions of resources, share of management control, profit and risk.

Whatever the source of finance under consideration marketing has a role to play in evaluating the appropriateness of that source as well as identifying it in the first place. A common requirement is that marketing proposals include a forecast of the payback period. Those responsible for developing these proposals are best placed to evaluate the compatibility between the market opportunity under consideration and the alternative modes of financing it. Of specific interest is the prospect of the investment payback period matching the repayment schedule. Enterprises which finance long term investments through short term sources of finance are either badly misinformed or have adopted a high risk strategy.

Risk bearing

In both the production and marketing of produce the possibility of incurring losses is always present. Physical risks include the distruction or deterioration of the produce through fire, excessive heat or cold, pests, floods, earthquakes etc. Market risks are those of adverse changes in the value of the produce between the processes of production and consumption. A change in consumer tastes can reduce the attractiveness of the produce and is, therefore, also a risk. All of these risks are borne by those organisations, companies and individuals.

Risk bearing is often a little understood aspect of marketing. For example, when making judgements as to whether a particular price is a 'fair price' the usual reference point is the producer or supplier's costs. However the risks borne are rarely taken into account by those passing judgement and yet, almost inevitably, there will be occasions when the risk taker incurs losses. Stocks will spoil, markets will fall, cheaper imports will enter the country, consumer tastes will change, and so on. These losses can only be observed if adequate surpluses were generated in previous periods. Risk bearing must be acknowledged as a cost since what is uncertain is not whether they will occur, but when they will occur.

Market intelligence

As for as is possible marketing decisions should be based on sound information. The process of collecting, interpreting, and disseminating information relevant to marketing decisions is known as market intelligence. The role of market intelligence is to reduce the level of risk in decision making. Through market intelligence the seller finds out what the customer needs and wants. The alternative is to find out through sales, or the lack of them. Marketing research helps establish what products are right for the market, which channels of distribution are most appropriate, how best to promote products and what prices are acceptable to the market. As with other marketing functions, intelligence gathering can be carried out by the seller or another party such as a government agency, the ministry of agriculture and food, or some other specialist organisation. What is important is that it is carried out.

Links between Agriculture and the Food Industry

The link between agriculture and food continually evolves. In primitive societies, the farmer and consumer were either the same family or close neighbours who bartered their products and services, but as societies develop other linkages are added. Commodity traders, processors, manufacturers who convert produce into food items and retailers, among others, are interposed between the producer and consumer. A more recently introduced link into the chain is the scientist. Scientists as breeders, plant biologists, nutritionists and chemists have made an immeasurable contribution to the development of agricultural production and food manufacture over the past 50 years. It would appear that we have passed through the age of machines in agriculture, and the age of chemicals, on to the age of biotechnology in agriculture.

Biotechnology has great potential for the developing countries since it is likely to be less capital intensive and more research and know-how intensive. Thus its benefits can flow faster into the poorer countries who do not have the capital. Therefore its impact could be faster, more widespread and more significant.

As the link between food and agriculture continues to evolve, we see the emergence of an agribusiness i.e. where agriculture and food become a continuum. Multinational companies like Cargill, Brooke Bond Liebig, and Del Monte are examples of vertically integrated organisations with links all the way through from agricultural production to retailing. There is a line of argument which says that it makes sense that those who are closest should the consumer should assess his/her needs and interpret them back to the primary producer.

As disposable incomes increase, the food industry will increase the quality and diversity of the products it produces. Food manufacturers will have particular expectations of agriculture as a supplier of their raw materials, including:

Quality

To build a profitable business, food manufacturers seek to establish a preference for their products by differentiating those products in some way which is meaningful to consumers. Then, in order to enable consumers to recognise the differentiated product, manufacturers brand that product. Manufacturers can then work on building consumer loyalty to these brands. Brand loyalty is normally only established by delivering high quality consistently. As disposable incomes rise, the market tends to develop more sophisticated needs and the quality of the raw material becomes even more critical. Where agriculture is seeking to serve a food industry, that itself is seeking to meet these more sophisticated needs and wants, it can expect to face increasing emphasis on quality. Equally well, agriculture can expect to share in the better return for innovative improvements in quality.

Cost

Next to quality will come cost. With an increased capability to search the world for raw materials, the food industry is able to find the lowest cost source for any given level of quality. For the food manufacturer, the country in which he/she manufactures, or markets, need no longer be the source of agricultural produce. Improved transportation and communications mean that the world is becoming his/her source of supply. This is a significant change in the competitive environment of agriculture which the farming community has to realise, because they have, hitherto, been largely concooned in their respective domestic markets.

Non-seasonality

Agricultural products were traditionally seasonal in their production and supply. Modern technology and husbandry practices mean that food manufacturers need not have their production schedules dictated by the seasons. Indeed, the capital intensive food industry cannot afford to incur the high costs of under utilising its capacity. This means that farmers will have to complete in terms of reducing seasonality or fitting into a pattern of social competitiveness.

Reliability

A manufacturer who has invested heavily in building up his brand will be very keen to get reliable supplies in terms of quality, timing and cost. Producers of agricultural produce will be increasingly judged on their reliability in all of these respects.

Processing

Ease of processing will become an increasingly important expectation of the food industry. Like all industries, reductions in the costs of capital equipment, wages and inventories are important objectives. For example, farmers who can deliver on the 'just-in-time' principle will contribute towards reducing a manufacturer's working capital and space requirements. Farmers who can do part of the secondary processing and/or performing functions such as the post harvest treatment of the crop or transporting will be adding another advantage. Crops that are specially bred or designed to facilitate processing (e.g. seedless fruits, featherless chickens, coffee beans without caffeine, low cholesterol meats) are another type of advantage that the food industry could expect from agriculture. In short, the competitive advantage will rest with those able to add most value and can differentiate what they are offering from that of other suppliers.

Product Differentiation

In competitive brand marketing, the food industry has to innovate continuously to create new products that are different from and superior to existing ones of their own or competitors. The scope of innovation has traditionally been at the processing stage. Whilst this will continue to be an important area for innovation, manufacturers will increasingly tend to look for innovative changes in the agricultural produce itself. This may be in terms of novel tastes, improved texture, more attractive shapes, etc.

Health Aspects

We have already said that in the more sophisticated food markets, healthy eating can become a priority among consumers. Therefore, farmers will have to consider the health connotations of what they choose to grow. There are two aspects of health to be taken into account. First, consumers may be interested in the food itself i.e. low fat, low/no sugar or low/no salt. It would be a mistake to think that health issues are confined to the more sophisticated food markets or to the wealthier segments of the community. Nutrition is important in all segments of the market. Even where the poor receive adequate amounts of food to fend off starvation, they are often malnourished. Thus farmers have to be concerned about the nutritional value of the produce they grow. Second, the consumer may be more, or equally, concerned about the food production methods i.e. the avoidance of chemicals like herbicides, pesticides etc. This may mean a change to the farmer's husbandry practices with implications for the costs of production. The consumer and the food industry will expect the farmer to produce without potentially dangerous chemicals, but at no extra cost to them. This will be another challenge for agriculture.

Agricultural and Food Marketing Enterprises

The principal component of any marketing system are the institutions and enterprises of which it is comprised. Three of the principal forms of enterprise to be found in developing countries are discussed in this section. These are: private companies, marketing boards and co-operatives.

Private Enterprise

Private enterprise has much to commend it, including a much higher level of financial independence from government than public enterprises. Moreover, private enterprise is able to adapt, rapidly, to changing circumstances and opportunities and is usually able to provide what consumers want at a lower cost than public enterprises. Abbott highlights several particular strengths of private enterprise, including:

Table 2. Strengths of private enterprise

Low operating costs	Nothing so concentrates the mind on cost control than ownership. The private entrepreneur has every motivation to contain costs since to do otherwise erodes his/her profit margin.
High level equipment utilisation	Since private enterprise has as its prime objective, profit, everything is done to maximise the use of capital equipment, and thereby lower unit costs e.g. concern is shown to keep factories operating at high levels of capacity utilisation, attempts are made to ensure that the firms' vehicles have economic return loads as well as outward loads etc.
Adaptability	Decision making within private enterprise tends to be quicker, because of the absence of a weighty bureaucracy, than in public enterprise equivalents.
Personal initiative	The entrepreneurial spirit is in evidence when an individual shows a willingness to accept calculated risks.
Rapid decision making	Decision making within private enterprise tends to be quicker, because of the absence of a weighty bureaucracy, than in public enterprise equivalents.
Independence of spirit	Entrepreneurs need a good deal of self confidence i.e. they must be prepared to back their own judgements rather than rely on the views and support of others. Moreover, it often takes a fair amount of time before market demand can be built up and new markets penetrated and hence the need for tenacity.
Willingness to work hard	There is a direct relationship between effort and the level of success in private enterprise. Rarely is the entrepreneur able to rely on others covering for him/her and no-one pursues potential business or seeks to solve management problems with as much vigour as the owner.
Relevant experience/expertise	Most successful private entrepreneurs have experience and/or expertise which others are willing and able to 'buy'. This could be, for example, the ability to judge the quality and quantity of meat a live animal will yield when slaughtered.

An understanding of agriculture	This of course, relates to agribusinesses and is essential to those seeking to do business with farmers (or fisherman). Knowing how crops are grown and mature and understanding the priorities of producers and the daily/seasonal pressures they face is invaluable in agribusiness.
Perishable products	This class of product is subject to rapid and extreme fluctuations in supply and demand, and therefore price, as well as considerable variation in quality, both at harvest time and subsequently, due to mechanical, pathological and/or physiological damage. A private company, with its ability to make quick decisions, in response to an ever changing environment and set of market conditions, is in a better position to prosper in the perishable produce market.
Livestock and meat	Abbott claims that the marketing of livestock and meat is dominated by private enterprise. He says that this is explained by the fact that direct decision making gives private enterprise the edge because of the need for skilled judgement in appraising quality and value when the product is so variable.
Combined purchase of produce and sales of farm inputs and consumer goods	Businesses serving rural customers often have to deal in small quantities of supplies and purchases and this requires a great deal of flexibility on the part of the enterprise. It is usually the smaller, private, enterprise which proves willing and able to conduct business in such a way.
Specialised activities in marketing	The willingness to invest in new, and therefore risky ventures or to invest in highly specialised activities is usually the province of the private sector.

Marketing Boards

Marketing boards are, in most instances a government agency and/or statutory organisation having the function of intervening in the marketing process, with a view to serving the cause of efficient and orderly marketing. Less frequently they are voluntary organisations established by farmers/ producers. Put another way, marketing boards tend to be born out of government policy rather than by consensus among commercial parties. This is especially true of Marketing boards in the tropics where their chief object is to improve the income of the smallholder, grower, and/or livestock farmer.

Marketing boards do not normally provide marketing services to large estates or plantations. Prior to the adoption of structural adjustment and market liberalisation in nearly all Marketing boards served as 'price stabilising boards'.

Another characteristic of marketing boards is their focus on durable products. Marketing boards are normally given authority for 'controlled' or 'scheduled crops'. In many countries fewer than 5 crops are controlled. These tend to be traditional crops like millet, sorghum, rice, maize, groundnuts and palm oil and 'colonial' crops such as cocoa, cotton, coffee, tea, tobacco and rubber. Some governments have opted for boards that control more than one crop. In some cases, the marketing board performs all of the marketing functions itself but in others it cooperates with private enterprise by, for example, hiring storage facilities or appointing local buying agents.

The effectiveness of a particular marketing board is often viewed in terms of three factors:

- Its contribution to orderly and efficient marketing
- The reduction in the capacity of intermediaries to manipulate margins at the expense of producers and consumers
- The generation of producer-oriented monopoly power

In many cases the establishment of a marketing board was a reaction to situations where middlemen and/or foreign buyers were perceived to hold monopsonistic power over producers. Hence the role of the marketing boards is frequently articulated as being one of organising producers into monopolistic agencies with real countervailing power; to reduce inefficiencies due to unwarranted competition, and duplication of effort between intermediaries. In theory at least, the marketing board contributes to orderly marketing by acting as an agent for improving marketing practices, as a market regulator and as a provider of facilitating services. For instance:

1. *Change agent:* Marketing boards can establish marketing practices and procedures for raw and/or processed products.
2. *Regulatory role*: Marketing boards may act as "watch-dogs" over agreed marketing practices and procedures e.g. credit arrangements, weights and measures, quality control etc.
3. *Facilitator:* Marketing boards may provide all or some of the facilitating services e.g. credit, market intelligence and risk management. The last of these usually takes the form of the

guaranteeing of prices. In the case of tree crops prices are announced in advance of harvest. Prices for annual crops are normally made known before planting or sowing.

The role of marketing boards in bringing about more efficient marketing is most often framed by policy makers in terms of modifying the market structure. That is, trying to make what is perceived to be an imperfect market structure more advantageous to producers. Of course, in doing so, account ought to be taken of the effect on both consumers and other players within the marketing system. This is not always done and the question is begged whether a market structure which is organised to the principal benefit of one particular set of players is anything other than imperfect to the others.

However, the argument in favour of giving producers real countervailing powers is strongest in situations in which the marketing system is characterised by a myriad of largely powerless producers and a relatively small number of powerfull intermediaries. In these circumstances, the price-makers are the middlemen and both producers and consumers are price-takers.

One particular way that a marketing board may act to modify an existing market structure is to rationalise the system in an attempt to reduce inefficiencies seen to be caused by unwarranted competition and duplication of effort between intermediaries. For example, there may be duplication of transport, storage and processing facilities to the extent that capacity utilisation cannot rise to economic levels without extremely high charges to compensate. Marketing boards may try to rationalise the system through, for example, a system of licenses.

Buying Operations

Marketing boards would normally buy at fixed prices. Each season or year, the government sets the price for scheduled crops. In the case of tree crops, this price is announced before harvest and before planting or sowing in the case of annual crops. It is subsequently kept at the same level for a period of time: typically about 6 months. These procedures give some security to producers.

Buying takes place at official buying points where there are either appropriate storage facilities for the produce, or transportation so that it can be moved before any significant deterioration in quality occurs. Clearly farmers are concerned that buying points should be conveniently located.

However, maintaining an extensive network of buying points adds substantially to a marketing board's operating costs and so the interests of the two parties often conflict. One compromise is for the marketing board to operate mobile buying teams to supplement permanent buying points.

In some countries the buying points are staffed by board employees, but the costs of running the buying points and the associated transportation costs can become too high and some governments seek alternative solutions such as transferring buying points to local co-operatives, and/or by appointing licensed buying agents (LBAs). It is common in anglophone West Africa to have co-operatives operating as LBAs, in competition with private traders who are also LBAs. The use of buying agents promises some degree of competition, which in view of the fixed prices, expresses itself in the secondary conditions, in particular better service.

Selling Operations

Some marketing boards, like grain boards, are concerned entirely with domestic consumer markets. These tend to be handling staple crops such as maize, millet and rice. Other boards are dealing exclusively with export markets and, therefore, industrial buyers. The two types of markets are quite different from one another and so therefore are the operations of the boards serving them. A distinction is sometimes drawn between these two types of board by referring to Food Marketing Boards (FMBs) and Export Marketing Boards (EMBs). Among the major differences is the position of governments with regard to them. First, governments have no control over demand in export markets whereas they can, and do, exert control over demand within the domestic market. Second, since governments have to take account of the interests of domestic consumers of staple crops, they sometimes instruct FMBs to adjust their marketing strategies to meet social and/or political rather than commercial objectives. The interests of consumers in export markets are of no direct concern to the government of the exporting nation.

Selling operations of EMBs

Some export markets are governed by commodity agreements such as the Sugar, Cocoa and International Coffee Agreement, but in the majority of cases they must operate within free or open markets where vigorous competition exists. EMBs tend to favour early sales. That is, they try to minimise the time period between buying and exports. This is sometimes termed a 'rapid evacuation' policy. It keeps storage and capital investment requirements to a minimum, since the burden of holding and financing stocks is carried by the recipient of the produce. Most EMBs practice 'forward selling' which as the term itself suggests, means signing sales contracts well in advance of delivery. Sometimes it means selling the crops well in advance of their being harvested, or sometimes even before they are planted.

The practice of 'export parity pricing' is prevalent among EMBs. This means that the producer price is calculated as a residual of the export price minus marketing costs. There is no particular motivation to minimise those marketing costs in such a system, since a major source of uncertainty for EMBs has always been the instability of prices in the open world markets. As EMBs cannot influence these prices, they tend to take the defensive approach of lobbying for low producer prices. In this way they hope to avoid a trading deficit when world prices fall.

Selling operations of FMBs

In many developing countries the FMB's selling price is set by government. Concern for the welfare of consumers often encourages governments to set low prices. This means the gross trading margin of an FMB is often small. The margin is invariably a source of conflict between FMBs and the government. In its desire to please both consumers and farmers, government will often suppress the profit margin and insist upon the FMB reducing its outgoings. The government usually has the upperhand but since it has to bear any deficit it is a hollow victory.

Consumers needs determine the timing of the release of stocks. Staple crops, usually have a fairly constant demand throughout the year, and FMBs have to bridge the usual, and considerable, interval between buying after harvest and staggered selling over the year. Stockholding is an important but expensive function of FMBs especially immediately after harvest when there is often insufficient storage space for the incoming produce. Conversely, as the stocks are slowly released FMB stores are under capacity for much of the year. A common objective of FMBs is basic 'food security' in times of shortage. This policy makes a lot of political sense but commercially it presents difficulties. Working capital is required for a longer period, and, if after all there is no shortage, the FMB is left with decaying stocks.

Nearly everywhere there is a 'dual marketing system', with a parallel market which allows farmers, traders and consumers to by-pass the FMB.

In some countries the parallel market is permitted by the government. Where FMBs have been given a monopoly, parallel markets become black markets, suppression of which has proved impossible. Indeed whether the parallel market is permitted or forbidden, the FMBs have to reckon with its competition.

Some boards do not fit easily into the two categories discussed so far. Produce such as groundnuts, sunflower seed oil and palm oil, have both domestic and export markets. Marketing boards handling these products have been mainly been established in countries where a surplus for export exists. These boards are normally classified as export boards. However, there is always the possibility that domestic demand will increase to the point where it absorbs the export surplus, at which point the board becomes a domestic marketing board.

On the whole the picture of marketing boards in the literature is a depressing one. They are largely portrayed as weak organisations which have achieved little success. There are however some outstanding success stories like the Zimbabwean Cotton Marketing Board (CMB) part of which is related below[a].

When setting out to evaluate the economic performance of marketing boards, it is all too easy to neglect to acknowledge that to a very great extent they are charged with achieving political as opposed to purely commercial objectives. Indeed policy makers often refer to them as 'instruments'. Those who readily identify the "mistakes" of marketing boards more often than not neglect to distinguish between those errors which could be rightly attributed to the boards' management, and therefore can be corrected by that management, and those lying outside the board itself. Some criticisms would best be addressed to those governing the activities of the board. For example marketing boards are frequently used, by governments, as instruments of national policy, including:

- the promotion of agricultural and rural development with social goals overriding commercial objectives
- as instruments of fiscal policy
- as a mechanism for containing urban wages through price restraint on staple foods
- as a device to encourage farmers to grow and sell more food and export crops, by pushing higher producer prices

- as a means of consolidating power by placing political appointees on to the Board.

Thus, many of the decisions and activities undertaken by marketing boards which adversely affect profits and cash flow are not the product of poor commercial judgement on the part of management but are attributable to individuals, outside of the boards themselves, who have goals that are entirely divorced from the efficient and effective operation of those boards.

Co-operatives in the Agriculture and Food Sectors

The co-operative enterprise has its origins in the 19th century and has become one of the most ubiquitous examples forms of business/economic enterprise. Co-operatives exist in all countries of the world and operate under diverse political systems: from communism to capitalism. The majority of these co-operatives are, through their national apex organisations, ultimately in membership of the International Co-operative Alliance (ICA), the representative world body of co-operatives of all types.

The motivation to form co-operatives has three particular aspects:

- the need for protection against exploitation by economic forces too strong for the individual to withstand alone
- the impulse for self-improvement by making the best use of often scarce resources
- the concern to secure the best possible return from whatever from of economic activity within which the individual engages whether as a producer, intermediary or consumer.

It is the belief that each of these aspirations can most advantageously be pursued and secured in concert with like-minded people that provides the stimulus to co-operative action. The underpinning principles with are those of self-help, voluntary participation, equity, democracy, and a common bond of common need and purpose. The cohesion of the group is maintained by ensuring that individual members cannot secure power or gain advantages at the expense of the others. Co-operatives reward participation in the co-operative venture rather than rewarding capital Self-interest is a primary motivator in co-operative enterprises, with economic gain being the primary objective. In these respects, co-operatives differ little from capitalistic enterprises; self-interest is simply pursued in a different way from the capitalist enterprise. Thus, the rate of interest paid on share capital is fixed and limited, and not subject to variation according to the amount of profit made. Secondly the use and distribution of surplus is restricted to one or more of the following purposes:

- allocation to reserves, where it becomes collectively-owned capital and is thereafter non-distributable
- for use on, or donation to, common-good, community project
- distribution to members in proportion to the trade each member has done with the co-operative. In other words, the distribution is made not in relation to capital held, but by declaring a bonus or dividend per cash unit of trade done

Structure and Organisation of Co-operatives

There are two principal forms of co-operative organisations: primary co-operatives and secondary co-operatives. The basic unit in the co-operative systems is the primary co-operative. A primary co-operative is one in which the shareholder are individuals; each of them having an equal share in its control.

1. ***Primary co-operatives***: In many cases, primary co-operatives will combine several functions e.g. an agricultural co-operative may provide consumer supplies to its members. Primary co-operatives may also own and run subsidiary enterprises related to their main functions, such as a consumer co-operative with its own manufacturing/processing or servicing business.
2. ***Secondary co-operatives***: While a primary co-operative has individual persons as members, a secondary (or federal) co-operative is one in which other co-operatives are the members. Apart from this basic difference the structure and organisation of both types follow a very similar pattern.

Control and Management of Primary Co-operatives

The control structure of co-operatives is made up of three tiers:

1. The General Meeting of Members
2. Management Committee
3. The Manager or Secretary

The General Meeting of Members makes policy and through this meeting members exercise control. In most countries there is a legal requirement to hold an Annual General Meeting which has the particular responsibilities of receiving and deciding upon an audited statement of account, deciding how any surplus shall be used and distributed, and of electing a committee.

The General Meeting of Members delegates the operational control of the co-operative to a management committee (or board of directors), which controls the works of the co-operative on behalf of the members. One member of this committee is elected chairman or president. A manager (or secretary) is appointed by the management committee as the chief administrative officer of the co-operative. He/she is responsible to the committee for the day-to-day control of the business. In small co-operatives he/she may be a member elected to do the work without pay.

Federal or Secondary Co-operatives

Secondary co-operatives (also variously described as "union" or "federal" co-operatives) can be organised for many different purposes. It is quite possible, and quite common, for a primary co-operative to be a member of several secondary co-operatives, depending on its needs and the local co-operative structure. Examples of secondary co-operative organisation would be:

- a local district union of 3/4 cotton marketing co-operatives to operate a ginning plant
- a federation of 2, or more, consumer co-operatives to operate a bakery
- a national union of agricultural co-operatives to manufacture fertiliser
- a national union of consumer co-operatives to organise wholesaling services and to manufacture merchandise
- a national federation of co-operatives to run a national bank.

Through the device of federation, co-operatives are able to organise very large-scale business operations at the national - or even international - level without detriment to the democratic control of the primary co-operatives by their own members. The secondary co-operative can, because of its larger volume of business or its wider representational base, undertake functions, provide services, and make representations, which would be beyond the capacity of all but the very largest primary co-operatives. Secondary co-operatives are a from of vertical integration providing the opportunity for economies of scale, scope for development and improved administration.

Secondary co-operatives can in turn form other secondary co-operatives - sometimes called tertiary co-operatives. In many countries there is one apex federation representative of all other co-operatives in the country and providing, at the national level, representative, advisory and professional services to the co-operative movement as a whole. These national co-operatives can then be affiliated to international organisations such as the International Co-operative Alliance.

Control and Management of Secondary Co-operatives

The control and management of secondary co-operatives is similar in form to that of primary co-operatives. The share holding members - the primary co-operatives - exercise policy control through the General Meeting and elect a management committee to act on their behalf.

The management committee in turn appoints a chief officer to manage the operation under its direction. The bye-laws of secondary co-operatives, as with the primaries, set down the organisational rules and procedures and are subject to the approval of the responsible local authority. The operating surplus of a secondary is also used and distributed following the same principles as a primary co-operative.

The federal co-operative can tend to become the masters of their member co-operatives rather than their servants. This situation can arise two reasons. Secondary co-operatives engaged in manufacture and trade can usually only operate efficiently given a high level of integration between their operations and those of their members. This requires a commensurate high level of discipline or some from of contractual compulsion.

Secondly, the operational size and volume of trade of secondaries can be such, compared to individual primaries, that there is a strong tendency for them to behave as the dominant partner in the relationship. It is a tendency that has to be monitored and, where necessary, checked. The secondary co-operative has as its chief obligation, the provision of services to member co-operatives.

The potential of co-operatives is immense. Co-operatives appear well suited to the economic, social and institutional needs of development in the rural economy. Co-operatives can provide the mechanism to organise and mobilise people for self-help action in providing the services they require as a farming and rural community. As self-administered rural institutions, co-operatives have the capacity to reflect, and to respond to the needs of their members; and, at the same time, to help foster attitudes of self-reliance and self-confidence within a framework of mutual aspirations and mutual action. In the delivery of services to their farmer-members they can provide an essential support to the development objectives of both the farmers themselves, and of national development policy.

As business organisations co-operatives also have the capacity to act directly as development agencies. In their steady accumulation of business assets, the expanding range of their services, the acquisition and use of management skills, the employment of staff, they are involved in a positive and measurable development function. Moreover, the flexibility of co-operative organisation, for example through the potential of secondary co-operatives, offers opportunities for collective action in the development of agro-industrial enterprise to help support and strengthen local initiatives, and to give a further boost to rural development. There are few countries where co-operatives are not recognised as potentially important agencies of development.

Weakness of Co-operatives

Unfortunately, the potential of co-operatives, and the extent of their development, has, in many cases, fallen for short of expectations. Low standards of performance, bad management, financial failure, corruption and misuse of funds, use of co-operatives for political ends, have been common features of co-operative enterprise in many countries. As a consequence, a great deal of understandable criticism has been levelled at the co-operative system, and many, including some members, have become cynical as to its ability to play an effective role in the development process. There are a number of problems which inhibit co-operative development and adversely affect performance, the more important of which are discussed below.

Realism of Objectives

Commitment and purpose are two important ingredients in motivation. Achievement of purpose is equally important. Objectives are expressions of purpose and expectation. To serve as motivators and guides to action they have to be attainable. The resources available have to be adequate to achievement of the objectives, and aspirations must be matched to ability. Neither members nor others should expect too much of co-operatives, including expecting them to expand too quickly. Most agricultural co-operatives in developing countries operate in commercial circumstances which any form of business enterprise would find difficult. Like their farmer-members, co-operatives have to operate in very marginal conditions. Their members are usually poor, often subsistence, farmers. High operating costs, low margins, relatively low turnovers, narrow stock inventories, seasonal trading patterns, exposure to the consequences of crop failure, high credit risk, fluctuating demand, are all familiar aspects of trading in such circumstances. Indeed, were it not so, it could be expected that private enterprise would have moved in to exploit a profitable market. It is not uncommon for co-operatives to be introduced to provide essential services because other agencies have either failed, or refused, to do so.

Expecting too much of co-operatives is one fault, expecting too much too quickly is another. The mistake is frequently made that once a co-operative appears to be reasonably well established, injection of loan capital from some external source will permit it to rapidly expand its services. Such hasty injection of loan capital can strain management resources, encourage unwise risk-taking, weaken financial judgement, lead to overstocked inventories and promote loss-making enterprise. Co-operatives ought to be allowed to develop at a pace commensurate with the ability of members to manage, control and finance the development. They should be permitted to expand steadily like any other successful business enterprise, finding the resources to do so largely from surpluses made in their own trading operations. Business capacity should not be strained, for example, to meet the objectives of a government development policy. Revolution rather than evolution, will only prove detrimental to both the viability of the co-operative and to the attainment of the policy objectives.

Conflict between Economic and Social Purposes

Economic success is basic to the achievement of co-operative purpose for, in the long run, unprofitable enterprises cannot be sustained. However, co-operatives are constrained in the extent to which they can mimic the objectives and practices of capitalist enterprise without abandoning the fundamental values of the co-operative movement. For example, in the pursuit of business growth there can be a strong temptation to weaken member control and concede greater control to professional management, to make the creation of profit a paramount consideration, and to ignore the concepts of equity and fair dealing. The creation of collectively-owned capital by reinvestment of profits (surplus) is a highly important and desirable practice, but has its disadvantages in that if the element of members' share capital as a proportion of the total capital structure becomes so insignificant that professional management can afford to ignore it and so ignore member control in making policy decisions. The outcome is an enterprise largely indistinguishable, except in name, from a capitalist enterprise.

Misuse of Co-operatives to Pursue Political Objectives

Attempts to divert the purpose and resources of co-operatives to the support of particular political objectives adversely affects the co-operative movement. Factional dissension among the group distracts it from the achievement of its economic objectives. Members' meetings can become political forums devoted to the advocacy of opposing views. In these circumstances many members can become disenchanted and lose interest, making it easy for a minority group to take control and to attempt to run the co-operative to serve its own ends.

Co-operative principles require that membership should not be assumed to imply either political commitment or obligation. Co-operative systems organised and tightly controlled by governments as instruments of state economic policy are rarely conducive to the development of democratically-controlled, member-owned co-operatives. They are created to serve the objectives of politicians and planners; objectives which may or may not coincide with those of the members who have little effective control of the enterprise.

Management

There has been a tendency to argue that a major cause of co-operative failure is the constraint imposed on the exercise of management skills and authority by the democratic nature of the enterprise. That being so, it is suggested that the authority of the General Meeting ought to be curtailed, leaving committees and managers to get on with the job of management. However, to do so would deny the purpose of the enterprise that being to enable people to run their own business. The solution lies in increasing and improving the level of member participation, not restricting it. Moreover, the standard of management within co-operatives is often inherently poor. As has already been said, co-operatives often come into being in markets and geographical areas considered as marginal in terms of profit potential by most other forms of commercial business enterprise. This being the case, the salaries, working conditions and work location that they are able to offer fail to attract top quality managers.

Selling Arrangements between Co-operatives and their Members

A principal policy question in co-operatives is the procedure to be used in selling members' produce. The alternatives are: outright purchase from members, or sale on commission.

Outright purchase

In this case members are paid for their produce, at prices fixed by the co-operative, at the time of delivery, and the co-operative takes title to the produce. The co-operative then resells the produce at the most advantageous terms it can secure. Profits made on the transaction will be used first to meet the operating expenses, any surplus balance being used or distributed by decision of the General Meeting. This approach requires the co-operative to have high levels of funds available. Since, in the case of seasonal crops, a lot of produce is being offered within the immediate post harvest period, a serious adverse cash flow situation can arise. This can be alleviated by a two-stage payment system whereby members are paid part of the sale price at the time of delivery, and the balance after the co-operative has resold the produce.

The main objection to outright purchase is that the co-operative carries all of the post harvest risks including: fall-off in demand, price fluctuation, reduction of produce value due to down-grading, deterioration giving rise to loss of quality and so value, failure of transport arrangements, spoilage, fire and theft. Some of these can be covered by insurance but most cannot. Generally, this method is only acceptable where the risks incurred are limited and can be reasonably well assessed. For example, where forward contracts have been negotiated. These risks being taken into account, outright purchase has the advantage of permitting the co-operative to add value to the crop and thereby add to the profits of its members. There are three principal ways in which a co-operative might add value to the commodity.

- Produce can be stored for sale at a later date when prices have improved

- Value can be added through primary processing of the crop. Cotton ginning is a good example of relatively simple and inexpensive process which is best done close to the fields. The value of baled ginned cotton, is normally considerably more than the sum of the value of the raw cotton plus the cost of ginning. And there is the additional value of the cotton seed
- Opportunity for adding value exists in the packing and presentation of the crop, or in the case of livestock, improvement in condition or quality before sale. Graded, washed and well packed fruits and vegetables can attract wider markets and premium prices.

There is a *caveat* which ought to be added. Whilst price advantages gained by adding value are of direct benefit both to the co-operative as a whole and to individual members, the additional investment needed to capture the additional return, can be prohibitive. Apart from the risks incurred, outright purchase, storage, packing, processing, transportation, marketing etc., require substantial financing. Moreover, these more complicated operations call for a high level of management skill and judgement, which is frequently a scarce resource.

Sale on Commission

This far simpler, virtually risk-free, operation leaves the co-operative as the producers' agent with no legal title to the goods. All attendant risks therefore remain with the individual producers. The co-operative collects produce from members and sells in the most advantageous markets. It then deducts a commission at a previously agreed rate from the sale price. The co-operative meets the cost of its expenses from its commission income. With the sale-on-commission system the co-operative avoids the need to finance crop buying and it minimises its risks. In addition, much simpler operating procedures are required and expenditure can be more accurately matched to anticipated income.

The main disadvantage of sale-on-commission is that neither the member nor the co-operative is able to exploit possible price improvement. Another is the possible delays in the producer receiving cash for his crop. No payment will be made by a co-operative until it has been paid by the customer. Apart from the time taken in an 'open market' for crops to be sold, it is by no means unknown for parastatal agencies to be dilatory in paying-out for produce received from co-operatives.

References

Abbott, J. (1987), *Marketing Enterprise For The Developing World*, Cambridge University Press, pp. 181–187.

Dinham, B. and Hines, C. (1983), *Agribusiness In Africa*, Earth Sources Research Ltd., London, pp. 115–119.

Gaedeke, R.M. and Tootelian, D.H. (1983) Marketing: *Principles And Applications*. West Publishing Company, Minnesota, p.11.

Kohls, R.L. and Uhl, J.N. (1990) *Marketing Of Agricultural Products, 6th edition, New York,* Macmillan Publishing Company pp. 18–21.

Selassie, H.G. (1995), *International Joint Venture Formation In The Agribusiness Sector*, Avebury, Aldershot.

2

Liberalization of Foodgrain Markets

As the economies of developing countries continued to deteriorate in the wake of both external and internal shocks, LDCs experienced increasing difficulties in obtaining external loans and became more dependent upon the International Monetary Fund and the World Bank for financial support. The World Bank observed that stabilisation measures implemented by LDCs had often fuelled inflation, e.g. increased borrowing, foreign exchange controls, the extension of price controls, increased import tariffs and control of imports through licensing. It was soon apparent that these remedies were ineffective and alternative solutions had to be found. International funding agencies began to insist that borrowing governments implement an economic structural adjustment programme (ESAP). The reform or liberalisation of markets including agricultural and food markets, has been a central element of ESAPs. In agrarian based economies, which would include almost all developing countries, reform of the agricultural and food markets has been the single most important component of ESAP.

Economic Structural Adjustment Programmes

The prolonged economic recession which began in the late 1970's accelerated the rapid deterioration in the economic condition of many developing nations in the form of an ever widening gap in their balance of payment accounts. Many LDC governments reacted by implementing 'curative' measures that actually fuelled inflation. These included increased borrowing, restricting access to foreign exchange, the extension of price controls, increased import tariffs and control of imports through licensing requirements. On occassion, such measures were pursued independently but on others they were part of IMF sponsored economic stabilisation programmes. By the early 1980's it was apparent that stabilisation measures on their own would not solve the problem and alternative approaches were sought. The World Bank was first to conceive an economic structural adjustment programme (ESAP). Soon most international funding agencies were insisting that borrowing governments implement ESAPs. A key element of all structural adjustment programmes has been the reform or liberalisation of markets, including agricultural and food markets.

The World Bank first applied the term *structural adjustment* to describe its programme of policy-based lending which began in the early 1980s. (The usual business of the World Bank is project-based lending). The objectives of structural adjustment programmes are not confined to restoring macro-economic balance but are also intended to stimulate economic growth. Structural adjustment involves improving the structure of production by allocating resources in accordance with their opportunity cost rather than on any other basis. The argument in support of this approach is that resource allocation efficiency is maximised, increasing the value of current output and improving the prospects for the rate of growth over time and avoiding the need for subsidies and taxes in support of the production structure.

In addition to increased allocational efficiency, structural adjustment is concerned with improvements in both operational and economic efficiency. Operational efficiency is increased when unit costs of production are minimised through efficient management and the adoption of the appropriate technology. Economic efficiency is the consequence of a high level of allocational efficiency plus a high level of operational efficiency and ensures that consumers' needs are satisfied at prices which reflect the minimum sustainable cost of production.

Whilst and improved structure of production, through increased allocational and operational efficiencies, is the central platform of structural adjustment, structural improvements in practice have also included diversification of the economy. That is, by broadening the base of economic activity the flexibility of the structure can be improved and so the economy is better able to withstand external shocks such as a fall in the world prices of certain commodities, adverse weather conditions that impact on particular crops or substantive changes to trading conditions applied by economic blocs like the European Union. The in-built inflexibility of LDC economies meant that the external shocks of the 1970s (e.g. quantum increases in oil prices, world wide recession and the collapse of prices for a wide-range of agricultural commodities traded on world markets) had a far greater impact than they would have had if these economies had a broader base.

The challenge for those charged with developing structural adjustment programmes is to find designs capable of simultaneously stimulating economic growth whilst protecting the population from excessive levels of hardship during the course of the programme. In addition, ESAPs endeavour to make the domestic economy more flexible so that it is better able to cope with changes in the world economy. Overall the objectives of ESAP are to:

- restore equilibrium in the balance of payments
- reduce the fiscal deficit to a manageable size
- bring down inflation
- improve long term employment prospects
- increase investment and
- lead to sustained growth.

In order to achieve these objectives policy makers must:

- identify an agenda of reforms in policies and institutions
- develop a sequence of measures which is technically consistent and politically sustainable and
- negotiate these policies between governments and the lending agencies.

The objectives themselves are at the macro-economic level. In most instances they are attempted through policies designed to manipulate supply rather than demand. These policies are directed towards reducing spending overall and redirecting demand towards domestically produced goods.

The World Bank began its structural adjustment programme in the early 1980s and viewed this as a short term divergence from the business of lending for projects. The expectation was that most countries that instigated structural adjustment programmes would have completed them within three years. Both the duration of individual programmes and the extent to which the Bank would be involved in policy-based lending were woefully miscalculated. Some structural adjustment programmes have extended over eight years or more, and many are in their second and third phases. Moreover, the World Bank's involvement in funding these programmes has continued to increase, rather than diminish, and the Bank has publicly conceded that structural adjustment is:

> "...not a one-shot effort...but reflects the need for macro-economic and sector policies to be continuously appraised and modified."

All of this points to the fact that a knowledgeable and respected institution like the World Bank is still learning from experience about the processes of structural adjustment. It should not therefore be surprising that other lending institutions, the international donor community, government ministers, policy makers and others involved in the development, implementation or administration of structural adjustment programmes are on a learning curve and as yet have no firm understanding of what constitutes 'best practice' with respect to ESAPs. This is hardly surprising since, as Scarborough and Kydd state:

> "In countries which have embarked on structural adjustment programmes the breadth of policy reforms, the pace of their implementation and the lack of data has been such that the possibilities for detailed analysis of policy options has been very limited."

Macro-economic Stabilisation

Whereas the World Bank has been the principal sponsor of ESAPs, which focus on the supply side of economies and impact indirectly on the demand side, the IMF has been the champion of economic stabilisation programmes. These operate to stimulate the demand side of the economy and have an indirect effect on the supply side of the economy. Stabilisation policies work to reduce a country's expenditure levels to match its current resources. They do not directly lead to higher growth rates but rather provide the economic stability necessary before increased growth can be a real prospect. Attwood explains the objectives of macro-economic reform, thus:

> "Macro-economic reform involves creating a conducive climate for investment, addressing the imbalances in public expenditure, reducing the fiscal deficit, increasing parastatal efficiency and appropriate debt management."

The policy instruments typically employed in the pursuit of economic stabilisation are:

- exchange rate policy
- fiscal policy and
- monetary policy

Thus, for example, export performance (demand) might be improved because the reforms could result in a devaluation of the local currency (exchange rate policy), and/or higher levels of investor confidence due to inflation being contained as government brings its own spending under control (fiscal policy) and interest rates are allowed to rise to their economic level (monetary policy).

During the earliest round of ESAPs there was a degree of specialisation between the IMF, which handled demand management measures, and the World Bank, which focused on supply management measures. However, during the 1980s the IMF and World Bank began to develop programmes for an

> "...orderly adjustment of both macro-economic and structural imbalances so as to foster economic growth while bringing about a balance of payments position that is sustainable in the medium term."

This move was initiated by the World Bank in 1980 when it introduced its Structural Adjustment Loan (SAL) programme. In the mid-1980s, the IMF launched its Structural Adjustment Facility and its Enhanced Structural Adjustment Facility (SAF and ESAF). These are longer term concessional lending instruments which closely resemble the policy-based lending facility of the World Bank. Countries wishing to take advantage of SAF or ESAF must first develop a Policy Framework Paper, covering the short to medium term and the government has to jointly agree this with the IMF and the World Bank.

There is much debate about the sequencing of reform measures. Discussion tends to centre around the ordering of structural adjustment vis stabilisation, liberalisation of domestic vis international markets, and the sequencing of measures intended to liberalise specific markets. There appears to be a widespread consensus that the liberalisation of domestic markets particularly those in the agricultural sector should precede attempts to liberalise international trade. It is argued that success in liberalising domestic markets would help establish a more robust and flexible economy, one better able to compete in international markets.

A central objective of structural adjustment has been to remove distortions in the economy resulting from government intervention and central control over markets. A secondary objective has been to improve the management of those activities that remain the responsibility of the state in the post-market liberalisation period.

Role of the State in Liberalised Markets

Countries undertaking market reforms have had to face the vexing question of what to do about government agricultural marketing institutions. Most of these parastatals have accumulated very large debts and deficits, and have, therefore, been an enormous drain on their treasuries. In some cases, the poor financial performance has been due to mismanagement and operational inefficiencies, but in others government has placed on the parastatal social responsibilities such as the management of the strategic grain reserve and developmental roles involving promoting the interests of smallholder farmers. In general, these social and developmental responsibilities cannot, by their nature, be profitable activities.Hence, the dilemma of governments seeking to reduce public expenditure, a large part of which has in the past been spent on financing agricultural marketing parastatals, but who cannot afford to allow social and developmental objectives to be abandoned in blind pursuit of liberalised markets. In some cases, governments have disbanded their agricultural marketing parastatals whilst others are seeking to restructure these marketing institutions.

The efficient allocation of resources is central to all structural adjustment programmes. This does not preclude the participation of the state in agricultural production and marketing activities. Historically, however, government involvement in many areas of economic activity has proved to be allocationally and operationally inefficient. It has been argued that government structures are not flexible enough to react to changing opportunity costs. Moreover, resources are often allocated on the basis of political considerations and operational inefficiencies arise from the absence of incentives. As a result, structural adjustment has also involved reducing the direct role of the state in marketing.

Strategies for Reforming Agricultural Marketing Parastatals

In broad terms, there are four alternative courses of action which can be taken with respect to the restructuring of agricultural marketing parastatals, these being:

- abolition of the parastatal, passing its essential functions to the private sector and/or non-government organisations (NGOs)
- transformation of the parastatal into a non-commercial organisation carrying strategic food reserves and stabilising prices
- transform the parastatal into a fully commercial enterprise, operating as a market oriented, profit making concern and
- transform the parastatal into a fully commercial enterprise, operating as a market oriented and profit making concern within a competitive marketplace.

The issue of privatisation is not a single option, but a policy direction which itself involves a range of possible options. These include:

1. ***Privatisation of management:*** That is, ownership of state assets to be retained by the state but management privatised by way of contract.
2. ***Privatisation of non-core assets and functions***: Another limited form of privatisation is the separation and privatisation of non-core assets and functions. This can involve the sale or lease of some manufacturing and distribution activities (e.g. collection, testing/grading, delivery to low density/rural markets), engaging in joint-ventures with private enterprise, etc.
3. ***Partial sale of equity***: It may be possible to sell part of the equity of a state owned company. Equity may be sold privately to particular groups, e.g. producers of the raw material, staff members of the enterprise, indigenous people etc. However, such a method of disposal is likely to be costly to the government since the equity would probably have to be sold at less than full market value. The government could opt to sell part of the equity through a stock exchange flotation.

Full privatisation This is the most difficult to achieve in the case of most agricultural marketing parastatals because they generally have social objectives that have overshadowed any commercial objectives. Potential investors may have little interest in being saddled with the residual social objectives, where these exist. Furthermore, their poor financial position makes them a high risk investment and their acquisition would immediately and adversely affect the share values and financial standing of their purchaser. Moreover, to fully privatise a parastatal would give a strong signal on future government policy with respect to public ownership generally.

The options discussed here need not be considered as being mutually exclusive. It is conceivable that several elements could be combined within a single reform programme. For instance, once the relevant market(s) have been liberalised, it would be possible to contract out the management of a parastatal and at the same time privatise non-core assest and/or functions whilst retaining public ownership of the parastatal. Other combinations are equally possible. The parastatal could be divided into several strategic business units (SBUs). For example, a Livestock & Meat Marketing Board could be divided into the following business units:

- procurement and fattening
- animal procurement advisory service
- auction management
- animal health and development advisory services
- abattoir services
- meat grading and packaging and
- meat marketing services.

In many senses each of these business units operates in different, although related, markets, as SBUs could each be a cost and profit centre. Once organised in this way they can be treated independently of one another. Thereafter, a number of opportunities arise due to the flexibility which such a major reorganisation would create. Depending upon circumstances, some could be privatised, some might be the object of joint ventures, other could remain in public ownership but their management be privatised in an attempt to increase efficiencies whilst reducing government expenditure and others might remain as government owned and managed public enterprises.

Obstacles to be Overcome in Commercialisation

An agricultural marketing board, or any other form of parastatal marketing organisation, can be commercialised and/or privatised. The process of commercialising an agricultural marketing parastatal involves management in establishing return-on-investment, turnover and profit targets. In order to achieve these targets the parastatal is expected to adopt a market orientation. In addition, the commercialised parastatal would be expected to raise funds on the financial markets of the private sector and pay interest at the market rate rather than rely upon public funding. Privatisation takes the additional step of transferring ownership of the parastatal out of the public sector and into the custody of private investors. The transfer of ownership can be effected in a number of ways. Shares in the enterprise can be floated on the stock exchange or can be sold to special interest groups such as management, employees or suppliers (e.g. a Dairy Marketing Board could be sold off to milk producers). The restructuring of a public enterprise involves two key preliminary steps:

- transformation into a genuinely viable commercial enterprise, earning a realistic return on capital and with a balance sheet fully aligned to trading results.
- transformation of the enterprise's special legal status into that of a normal commercial institution with no special trading rights or privileges conferred by statute.

An initial step might be to appoint a Board of Directors as Zimbabwe did with its four agricultural marketing boards. (i.e. the Dairy Marketing Board, Cotton Marketing Board, Grain Marketing Board and Cold Storage Commission). These independent directors can then be asked to exercise their collective judgement as to how the enterprise can be made more efficient and better meet customer needs. There then has to be a significant degree of deregulation of the markets which these enterprises serve, with more autonomy for the marketing boards, especially in pricing. Another priority is to deal with the accumulated debts of parastatals, where these exist.

Before the processes of commercialisation or privatisation can commence there are usually a series of obstacles to be overcome.

Accumulated Deficits

Boards which are in a relatively strong financial position are far better placed to undertake restructuring than those carrying large, accumulated deficits. The dilemma faced by policy makers is to decide how to handle these deficits. If they are simply written off by the government then this provides the parastatal with 'free' capital, an unfair advantage over potential competitors and provides no evidences of the organisation's future viability to prospective investors. Even where there is no intention to privatise the parastatal, the provision of free capital runs contrary to the spirit of transforming an organisation into a fully commercialised enterprise.

Source of Finance

Many marketing boards carry large deficits and need fresh injections of cash. However, the magnitude of there deficits usually discourages commercial lending institutions from supporting them. International donor agencies are a possible alternative source of funds but their rules of operation often prevent them from providing concessionary loans to the private sector, even recently privatised parastatals; these could however be available to a commercialised parastatal.

Tax Liabilities

Potential investors would be concerned about the liability of a restructured marketing board to corporation tax. A commercial company which makes losses over a period of time can usually offset these against profits made in other years, for the purposes of assessing its tax liabilities. How governments treat the losses made by marketing boards prior to privatisation will have a significant bearing on how they are perceived by potential investors. The magnitude of the losses are such that if the accumulated losses were written off against future profits, it might be a very long time before the enterprise paid any tax. On the other hand, if no allowance is made for these losses then the marketing board constitutes a most unattractive investment.

Developmental Roles

Marketing boards, and other types of parastatals, frequently have development roles, that is, activities and projects whose aim is to lead to the long term development of a sector of agriculture, a section of the population and/or a region of the country. For instance, a parastatal may be charged with helping convert subsistence farmers into cash crop farmers, or helping small scale farmers penetrate a dairy market dominated by large, commercial farms. Problems arise when the board has development responsibilities but no provision has been made for an adequate system of financing. In principle, the problem is easily solved in the case of commercialised, but not privatised, parastatals. That is, non-viable development activities should be separately identified and financed by government. The difficulty is that the accounting systems used by many parastatals are simply not capable of accurately separating out commercial from non-commercial activities or their attendant costs. Where the intention is to privatise the parastatal it is unlikely that prospective stakeholders would accept a development role unless the particular activities could be carried out at a profit. Even then, private investors would usually want to reserve the right to withdraw from an activity, if they identify alternative opportunities which offer a higher return on investment.

Another complication is the fact that governments tend to have wide-ranging political, social and economic objectives, many of which are likely to transend some of the aims of parastatals restructuring (e.g. indigenisation of an economy, drought recovery, etc.). This makes it difficult to deal with the restructuring or reform of an agricultural marketing parastatal in isolation from an array of confounding development issues.

Legislation

The restructuring or reform of marketing boards and other parastatals will invariably involve the amendment or repeal of a substantial body of legislation relating to a board and the products for which it has been responsible. Rarely is this a simple matter of withdrawing the prevailing legislation. In the case of agricultural products, legislation relates to a number of different issues economic, public health, research funding, etc. Whilst the processes of deregulation may focus on the economic and marketing issues, the public health measures cannot be swept away along with defunct rules and practices which related to controlled markets. Rather, it is likely that public health aspects will have to be reinforced and strengthened since deregulation often means that the number of market participants will increase in number. Similarly, the need for industry-wide funding of research and the provision of industry-wide services remains unaltered.

It is therefore necessary to review the whole corpus of legislation which directly or indirectly relates to the marketing board concerned, to decide the parts which can be repealed, those which should be retained and those which need to be amended. This process might involve the Ministry, the Office of the Attorney General, Parliament, a Cabinet Committee and then, perhaps, the Head of State. Each of these participants is likely to have a wide-ranging portfolio and agriculture is but one of them. Not surprisingly, therefore, effecting the necessary regulatory changes can take a considerable time, even when the government and the management of the parastatals are wholly committed to change. All too often, the time scale for legislative changes is underestimated by those who develop restructuring programmes.

The legislative issue is not of itself a major constraint, but the volume of work involved takes time. As in all aspects of government business, the passage of new legislation through the system will have to take its place with other new legislation before it can be passed into the law of the land. This inevitably affects the rate of progress of the restructuring of any government parastatal.

Dealing with Accumulated Deficits

Government capital investment in agricultural marketing parastatals often takes the form of irredeemable interest-bearing loans. The interest rate is usually highly favourable to the parastatal, compared to the rates prevailing on the open money markets. In most instances the government does not actually receive the interest due on parastatal deficits. Converting debt to equity improves the look of the balance sheet since debts must be repaid and at a specific point in time whilst equity belongs to the owners (i.e. shareholders) of the business and they only receive payment whenever the enterprise makes sufficient profits to pay out a dividend. A highly geared enterprise (i.e. one whose capital is largely comprised of debt borrowing) is generally perceived to be more of a financial risk for investors than one which largely depends on shareholder funds and retained earnings to finance its operations.

Where governments of developing countries have considered exchanging their outstanding loans to parastatals for equity, it is often solely with an eye to reducing or eliminating the interest payments of a beleaguered marketing board. This, however, is unsatisfactory where the enterprise is supposed to be of a commercial character. If little or not return is paid on the capital sums outstanding to government then this treats capital as a 'free good', which it is not. This is neither good in the campaign to engender 'commercial attitudes' within a former parastatal nor in the context of creating a 'level playing field' for emerging competitors (where these exist) who do not have access to free capital. Where governments provide 'free capital' to privatised parastatals it is an admission of failure to run the business as a commercial concern. Since equity capital carries higher risks than loan funds, over several years shares should earn a higher return, otherwise no equity investment would be made.

There may however be a case for writing off capital investments which were based on wrong assumptions or wrong policies in earlier years. For instance, investments may have been made in establishing poultry enterprises in rural years areas in anticipation of the electrification of those districts or villages. If for whatever reasons, that electrification has not taken place, those production units will be unviable and it would be best to write them off. A debt-equity swap would, at best, be self-delusion and at worst, would be concealment of a futile kind.

Encouraging Private Sector Involvement

In many cases the donors or lending institutions have made market liberalisation and encouragement for increased private sector participation part of the conditions attached to their structural adjustment loans. In some instances, the reform or abolition of a loss-making agricultural marketing parastatal has also been part of the agreement between the lender and the loan recipient.

Experience of implementing these reforms has been varied. In some cases, private sector participation has developed very quickly whilst in others, the process has been extremely slow. Much depends upon the stage of development of the private sector at the time market liberalisation measures are launched. In some countries, strong 'informal' or parallel marketing channels have co-existed with the formal channel throughout the period of administered pricing and "controlled" marketing of agricultural products. When these markets are deregulated the process of transferring 'informal' marketing functionaries into the formal sector tends to occur quickly. The entrepreneurs already exist and they have accumulated a certain amount of capital from their past marketing activities. On the other hand, where governments have been successful in suppressing informal marketing activities, the experience and culture of entrepreneurship and risk-taking may have been lost, or at least significantly diminished. In these cases, private sector participation in deregulated markets is a much slower process.

The Grain Market Board (GMB) held both a monopsony and a monopoly with respect to the staple food of the country, white maize. The GMB supplied large-scale millers located in the urban centres of Zimbabwe. A parallel channel did exist with unlicensed traders buying and selling grain and small-scale millers, using hammer mill technology, providing a service to customers. However, this illegal trade was conducted on a very small scale and the bulk of the grain passed through the formal single channel comprising the GMB and urban millers who used industrial roller milling technology.

The Zimbabwean Government paid subsidies to the GMB both in order to keep consumer prices down and to ensure a steady supply of the country's staple food. Prices to the farmers for maize were determined by the government rather than by market forces. Whilst in years of normal rains Zimbabwe can grow enough grain to feed itself and can even export white maize, the total amount of grain handled by the GMB was showing a discernible fall as the large-scale commercial farmers began to grow alternative and more profitable crops.

As trade in maize was deregulated, the number of custom millers increased substantially. (Custom millers mill grain which the customer brings to them. They do not provide the grain). A number of production millers also came into being. Production millers buy, mill, bag and brand maize grain using hammer mill technology. Market liberalisation brought many more players into the marketing system, giving growers a choice of market for their product. There is now more competition for the farmers' maize. Consumers have also benefited. They now have a wider range of products from which to choose. Price sensitive consumers can now buy grain direct from growers, or the GMB, and take it to a custom miller. The straight-milled maize is the cheapest of the options, available. Alternatively the consumer can buy the more refined product of the production miller who dehulls the grain before milling. This is more expensive than straight-run meal but not so expensive as the highly refined meal produced by roller mill technology.

The increase in the level of competition has also manifested itself in the form of marketing promotions. Prior to market liberalisation the active promotion of maize meal was virtually unheard of in Zimbabwe. The large-scale millers were assured of a sizeable demand for their product. After all, it is the staple food of the country and they were the sole suppliers. In the post-liberalisation period advertising expenditure on roller meal brands has increased substantially and in-store promotions are increasingly common.

Impediments to Private Sector Participation

Initially, the World Bank attached great importance to increasing the efficiency of parastatals as a means of bringing about market reforms. After all, agricultural marketing parastatals typically represented a sizeable burden upon the governments of LDCs and were characteristically inefficient.

However, in due course the Bank began to see greater potential in the liberalisation of markets and the privatisation of marketing institutions. Marketing reforms have tended to follow a pattern. In most cases, the first step has been to enfranchise private traders and actively encourage them to participate in markets from which they were previously excluded. The next step has been the removal of price controls. Marketing subsidies and parastatal subventions are then removed before the final act of market reform, i.e. the lifting of controls on international trade in agricultural products.

Scarborough & Kydd suggest that the following are necessary in order to promote an efficient private sector:

- encouraging and policing competition
- promoting the wider availability of information
- maintaining appropriate quality standards
- strengthening the legal system to ensure enforceability of contracts
- providing an adequate transport infrastructure
- undertaking spatial planning to allow appropriate access by marketers and consumers to central places
- supporting the development of credit
- research into the technological problems in storage and processing experienced by the private sector (which will be likely to be operating on a smaller scale, and facing different relative factor prices than parastatals) and
- training in technical and managerial aspects of marketing, processing and storage.

Prior to liberalisation government actions were perhaps the most significant impediments to private sector trade development. Although market liberalisation has, in most instances, changed government policy towards the private sector, the relationship between the state and commercial enterprises remains critical in determining the likelihood of success of any privatisation initiatives. The state effectively sets the rules of the game within which the private sector operates, and if these rules are too restrictive, or if they change too frequently, then private sector participation may well be inhibited. Here, consideration is given to some of the potential impediments to private sector participation in agricultural markets. As shall be seen, many, but not all, of these impediments are within the direct control of government.

Policy Implementation

Even when government is committed to encouraging private sector participation in agricultural marketing, the bureaucracy which implements policy changes may be less committed. Years of market controls and licensing may have created large vested interests in an underpaid civil service, which may see the spectre of retrenchment in deregulation measures. It is therefore not impossible that government policy is undermined at the implementation stage.

In the past, many LDC governments have been openly hostile to the private sector and so any trust established between the two parties is likely to be fragile. On the side of the private sector, bureaucratic delays will almost certainly be interpreted as an absence of real commitment, on the part of government, to free market economics and a suspicion that market liberalisation measures might be reversed at any time.

Licensing

A common post-liberalisation regulation is the requirement for traders to be licensed. This has been a feature in Malawi, Mali, Senegal, Uganda, Zambia and Zimbabwe, among others. Licences allow some residual control over the private sector; licences can be refused to traders suspected of unfair trading practices and can enable the state to collect marketing information. In some cases, the administration of trader licensing has presented prospective traders with few difficulties. In Malawi, for example, merely have to pay a registration fee and provide evidence of ownership of a seized weighing scale. In contrast, Uganda has used licence fees as an overt tax on traders; and in Senegal, traders dealing in more than 200 kg of grain have to obtain a licence, and an authorisation to collect a specific product. Authorisation is only given after the trader has shown proof of a minimum balance in his/her bank account. The administration of licences has been inconsistent. Sometimes the bureaucracy involved in applying for licences imposes substantial costs on traders in time, bribes and fees. At the beginning of a liberalisation process, licensing regulations may not be well publicised, or may change, leading to uncertainty among traders.

Quantitative Restrictions

Traders are sometimes limited to transporting loads of a specified amount. This can impose costs on traders, as there may be economies of scale, which increase the efficiency of the marketing system. Quantitative restrictions can be imposed in a variety of ways. For example in Zimbabwe, even though a maize marketing liberalisation programme had been implemented, the Grain Marketing Board (GMB) refused to sell maize in quantities less than 50 kgs. This effectively prevented very small traders, without transport, from participating in the maize marketing system. Even if the prospective Zimbabwean trader can afford to hire a truck, he/she probably will not be able to purchase enough grain to make up an economic load. Chisvo et al put the problem into perspective when they state:

> "Buying enough maize from the GMB to fill a 5-tonne truck requires almost twice the annual income of the average Zimbabwean...Those actually involved in grain trading, with few exceptions, reported that their only source of working capital was their own savings...Those who can capture scale economies using their own cash are relatively wealthy traders."

Rigid and Uncertain Regulations

Regulations, rather than direct intervention measures, can be the principal instrument of government policy. If regulations are too rigid then parallel or 'informal' markets may re-emerge. This has happened in both Senegal and Uganda. Moreover, access to the legal system, where this becomes necessary, has to be both quick and inexpensive.

The uncertainty surrounding the regulations which are to apply to the private sector can also prove an impediment to private sector participation in agricultural marketing. During the early stages of market liberalisation, changes to regulations may occur fairly frequently and this unsettles would-be investors in the agricultural marketing system. This can happen because the impact of liberalisation is difficult to predict, and as government monitors the process it may feel it necessary to tighten or relax the regulations surrounding the private sector as time passes.

Price Regulation

In order to retain a measure of control over food prices some governments have continued to impose price regulations in the post-market liberalisation era. These frequently take the form of floor and ceiling prices. These can have a substantial impact upon private trade. Private traders require a return on the resources which they invest in the marketing process. They will concentrate on those areas where marketing activities are most profitable, usually collection of grain near urban consuming centres, and ignore more distant markets where the information is poorer. This is certainly what happened in Malawi and Tanzania. Where government set narrow price bands within which the private sector must operate, these patterns are reinforced.

Smith recommends that government seeks to reduce the number and breadth of regulations to a minimum consistent with orderly production and marketing. He cites the example of the frequent requirement to obtain licences to trade and the need for separate permits for each commodity or groups of commodities in which traders wish to deal. Very often it is necessary to register the premises within which trade is to take place, and there may be restrictions on the quantities which can be traded and the times and forms of trading (e.g. wholesaling or retailing). Very often the reasons for many of these regulations have long since forgotten but traders remain on the statutes of the country. There are thus many instances where a simplification and reduction of regulations, and their consistent application, would remove a deterrent to private sector activities.

Public and Private Sector Competition

In many countries, liberalisation has allowed for the co-existence of public sector agricultural marketing institutions (i.e. parastatals and/or government supported co-ops). The public sector institutions may perform a market stabilising function, as say a buyer-of-last-resort, they may be the conduit through which food aid is released on to the market or they may manage the Strategic Grain Reserve.

Problems arise when public sector bodies benefit from support which is not extended to the private sector. For example, subsidies on inputs, such as fuel, may only be available through public sector agents. Similarly, public and private organisations may not have equal access to credit and finance. Many state bodies have historically had easy access to finance, often at subsidised terms, and without regard to the profitability of their operations. This has two negative effects on the private sector: it reduces the public sector's costs of operations, giving it an unfair competitive advantage, and it reduces the capital available for investment in private sector agricultural marketing operations. Tanzania's private sector became the dominant grain marketing force in 1990 when it was decided that parastatals and co-ops should only have access to credit on commercial terms. Neither of these parties was considered a commercially viable agency so they could not obtain the funds they needed to carry on their marketing activities at the same level as in the past.

Public sector marketing organisations have often enjoyed preferential access to foreign exchange. Input and output markets are often interlinked through credit arrangements. Where farmers rely on credit from marketing agents to buy imported inputs, private traders who have difficulty in acquiring these imports, may be disadvantaged in dealings over output trading. They may also have difficulty in obtaining access to fuel and transport, spare parts in particular, if they cannot compete for foreign exchange. In many countries, exporters are prohibited from retaining foreign exchange earnings but must convert it to local currency at official exchange rates. This has sometimes led to flourishing cross-border parallel markets. This caused the Ghanaian government to institute a foreign exchange retention system, in the post-liberalisation period, whereby cocoa exporters could keep a proportion of the foreign exchange they earned, thereby reducing the incentive for smuggling. A similar scheme operates in Tanzania where 35% of export earnings can be retained to purchase production inputs.

Inadequate Infrastructure

A major problem for traders in many countries is the low investment in infrastructure. In Africa, poor roads and inadequate storage facilities contribute to high marketing costs and commodity losses. In Uganda, some traders claim that the number of trips they can make have decreased to a fifth of what they could make ten years ago because of large increases in transport times and costs. Most small traders rely on public transport, or hiring space on trucks. Access to transport and ownership of trucks is often the basis for local monopolies in the transport sector, which increase traders' costs. The problem is all the more acute where foreign exchange shortages compel government to restrict the import of trucks and spare parts.

Adequate storage facilities can reduce crop losses and thereby food marketing costs. These often tend to be lacking in local markets, where a high proportion of grain trading takes place. Local market organisation is often the responsibility of local government. In some instances, market fees may be inadequate to maintain the market structure, or may be diverted to other areas of government expenditure. The absence of storage facilities also discourages traders from holding grain over time. This in turn prevents the development of private sector arbitrage which can moderate seasonal price fluctuations.

Credit Constraints

Most small traders have restricted access to formal credit markets, and tend to borrow from informal sources. These can be extremely expensive sources of funds and often lock the borrower into a never-ending cycle of debt. Commercial lending institutions are neither used to, nor motivated to lend to small-scale marketing enterprises. Financial institutions prefer to lend in large amounts to large, established enterprises operating in high profit markets. Complete liberalisation of financial markets can actually lead to a shortage of capital for investment in agriculture since it tends to be a low profit sector of the economy.

Lack of access to credit and finance prevents small traders from expanding. The quantities of commodities bought and sold, the amounts stored and transported and the ability to exploit economies of scale are all restricted by poor access to credit.

Problems are bound to arise in the agricultural marketing systems of countries which have implemented market liberalisation programmes if the financial system cannot be reoriented to service a large number of small private sector marketing agents as opposed to a few very large public sector organisations.

Marketing Information

With long established private sector marketing systems, informal networks usually exist to provide traders with information on market conditions, prices and the credit-worthiness of other parties. Conversely, where suppression of the private sector has been effective, in the past, the marketing information system tends to be underdeveloped. This was the case in Malawi where traders remained unaware of price differentials between markets and were unable to exploit opportunities for profitable market arbitrage which would also have increased market efficiency. Where there have in the past been strong parallel markets, information systems may exist but access to them may not be extended to post-liberalisation market entrants.

Impact of Macro-economic Environment on Private Traders

So far in this chapter consideration has only been given to impediments to private sector investment in newly liberalised markets that operate at the micro-economic level but there are also macro-economic aspects of the problem. In particular, private sector participation is affected by exchange rate policy, economic stabilisation measures and the level and nature of economic aid.

Exchange Rate Policy

To improve the balance of payments, exchange rate devaluation is usually implemented. Devaluation increases the cost of imports, and with it many agricultural marketing costs are increased e.g. packaging/bags, fuel and transport costs, agricultural inputs and imported foods. Where the private sector is involved in the marketing of imports, government needs to acknowledge where price increases are due to devaluation rather than the result of privatisation. Equally, where the private sector is operating under controlled prices, the allowance for marketing margins accommodates increases in cost due to devaluation.

Domestic Stabilisation Policy

Stabilisation programmes usually mean improved budgetary control through stricter monetary control, reductions in expenditures and tax increases. Each of these is likely to impact on both traders' costs and revenues. Austerity measures invariably reduce demand for most products and at the same time traders can find that they result in substantial increases in interest rates, licence fees, market fees, sales taxes, etc. The danger is that traders can be driven back into informal markets where some of these costs can be avoided.

Economic Aid

Structural adjustment programmes normally attract an increased inflow of aid from donors wishing to support the reforms. Much of the additional aid will be in the form of foreign exchange but some will be food aid or fertilizer aid. The way in which this aid is distributed can impact seriously on the development of the private sector. For instance, the private sector needs access to foreign exchange on the same basis as the public sector and co-operative agencies. Food and fertilizer aid distribution can have a more disruptive effect on private trade if it is not carefully timed. An influx of cheap food can reduce prices and cause considerable losses for private traders who have stored food grains. A system of distributing aid through auctions has worked well in Somalia and could work elsewhere.

Governments have to be careful in accepting aid projects which compete with potential private sector initiatives. These might be, for example, projects to build processing plants or storage facilities. Government has to be clear where it wants to encourage private sector investment. Governments can seek assistance from international donors on behalf of the private sector. Help can be sought to improve transport facilities or access to credit, or to fund training in technology, technical competence or business skills.

Government Action to Improve Private Sector Performance

There are a number of areas in which government can lend support to the development and expansion of the private sector.

Clear Policy Statements

The state has to make clear its view of the role of the private sector in agricultural marketing system, i.e. it must have clear objectives with the respect to the marketing system. It is not enough to simply lift existing restrictions on the functions of the private sector. A positive role should be developed for it. The government's policy has to be announced in unambiguous public statements, which make clear the extent of market deregulation.

It is equally important that the extent of the future involvement of the public sector in the agricultural marketing system be clearly articulated. If the public sector is allowed to enter the market, it must be known under what conditions, at what prices, and for what purpose this will occur. If a Strategic Grain Reserve is held then the trigger for grain release must be known to the private sector. Otherwise, uncertainty about the likely effect of government policy on market conditions will prevail.

Regulatory Framework

All to often in structural adjustment programmes the need for supporting legislation is either under-estimated or neglected altogether. Government can provide a set of stable regulations which would:

- clearly lay out the rights and obligations of consumers, producers and traders, underpinned by a carefully constructed contract law
- establish an acceptable system of weights and measures and perhaps a product grading system

- clearly specify the kind of exchanges that are legal, between which parties, and the constraints placed upon the place of exchange and
- the designation of an inspectorate to enforce the rules and regulations.

Training

Many of the entrepreneurs who come into a liberalised market will lack the basic knowledge and skills to ensure that their businesses can survive the periodic adverse economic conditions. Some may be technically proficient but know little about financial management or marketing; others may possess the basic entrepreneurial skills but have very limited levels of technical know-how. If government wishes to establish a private sector capable of growth and development then training programmes will be required. This does not necessarily mean that government has to provide or even fund the training. It may mean, instead, that government facilitates the training.

Infrastructure

Investment in roads, market places and storage facilities can reduce marketing costs and thereby increase levels of commercialisation. This might involve investing in new facilities and/or rehabilitating existing facilities. Where public sector marketing boards have under-utilised capacity, particularly in storage, this should be made available to the private sector, either through divestiture or through leasing arrangements.

Information Systems

The provision of market intelligence removes a barrier to market entry. Larger and longer established organisations, including marketing parastatals, have usually developed intelligence gathering systems over time. They have little motivation to share their information with smaller enterprises. In fact, to do so would lose them an important competitive advantage. This being the case, a government sponsored marketing information system would improve market transparency and help small-scale enterprises compete more effectively with larger organisations.

Transport

Transport shortages are a common complaint among private sector agribusinesses. This is especially true of small-scale businesses located in rural areas. Perhaps the best government can do is to give transportation a high priority within its foreign exchange allocation so that more vehicles, spares and fuel can be brought into the country. Emphasis should be placed on increasing the number of vehicles in the possession of the private sector.

Credit

Since poor access to credit and other forms of finance are a major obstacle to the private sector, especially the small-scale operator, government may have no choice but to get involved in establishing appropriate lending schemes. These may be financed from the Treasury, through special funds provided by international donors or through savings and credit schemes initiated by the private sector itself. Alternatively, government can encourage commercial banks to set special loan schemes targeted at various sub-sectors of agriculture and agribusiness. Due to the inherent risks in agriculture, and in dealing with small businesses with very limited resources, lending institutions will probably require some form of government underwriting of these schemes.

Liberalizing South Asian Foodgrains Markets

South Asia is home to the largest concentration of poor and undernourished people in the world, so food security—especially in basic staples such as wheat, rice, and corn—continues to be a major concern. With both persistent and re-emerging food price inflation reaching new heights in 2007–08 in global markets, South Asia saw sharp inflation—between 50 and 100 percent—in basic staples in Bangladesh, Pakistan, and Sri Lanka. These drastic price spikes drew comprehensive policy responses from the governments of these countries, addressing both supply and demand for foodgrains. India, the largest economy in the region, reacted by banning exports of common rice, wheat, and corn, as well as suspending these commodities from futures trading, to ensure comfortable supplies in the domestic market at affordable prices. India also launched a National Food Security Mission in 2007 and announced a special agricultural package (Rashtriya Krishi Vikas Yojana) of roughly US$6 billion to rejuvenate its agriculture.

Today, South Asian countries want a greater degree of self-sufficiency; reliance on trade to achieve food security is being questioned by critics. Against this backdrop, a recent book published for the International Food

Policy Research Institute (IFPRI) by Oxford University Press, Liberalizing Foodgrains Markets: Experiences, Impact, and Lessons from South Asia, studies the nature of reforms in foodgrains markets (both within-border and at-border reforms), their evolution, and their effects on food economy in general and food security in particular.

Due to their common colonial heritage and the universality of the problem, Bangladesh, India, Nepal, Pakistan, and Sri Lanka share similarities in their food policies. During the initial phases of development, the governments of these countries were slow to shift from a pro-industry bias, but the eventual adoption of agricultural technology brought about measurable improvements in food production and availability. India and Pakistan have emerged as food surplus states, Sri Lanka and Bangladesh have seen their food deficits decline, and Nepal currently faces only a slight food deficit.

As the situation of self-sufficiency in food production improved in South Asia, policymakers began to realize the drawbacks of interventionist policies, including spiralling costs, consistent inefficiencies, leakages, and corruption in the food management system. Hence, food policy changes—some of them in response to international commitments and pressures from donors—emerged. Governments of the region began to liberalize their food policies, with Sri Lanka leading the way followed by Bangladesh, India, Nepal, and Pakistan.

Public Procurement, Storage, and Distribution

Parastatal agencies such as the Food Corporation of India, the Nepal Food Corporation, the Pakistan Agricultural Storage and Supplies Corporations, and specialized state agencies or departments within the government—including India's Commission for Costs and Prices and Pakistan's Agricultural Prices Commission—worked on a set of food security objectives: to protect farmers' incentives; to stabilize prices, thereby reducing market risks for farmers; and to provide grains to consumers at subsidized prices. These elaborate systems of public intervention—from procurement to storage and distribution—have been scaled back in Bangladesh, Nepal, and Sri Lanka. In 1979, Sri Lanka replaced its system of universal rationing with a food stamps program while providing subsidies to consumers. In the late 1980s, Bangladesh dissolved its rationing system and limited public distribution to include only military, police, and public servants. In 1998,

Nepal, too, began restructuring its distribution-related food policies. It downsized the Nepal Food Corporation and started placing greater reliance on open market operations. India and Pakistan, however, continue to use elaborate systems of public intervention in different aspects of grain marketing.

Domestic Trade Policies

Elaborate control over domestic trade was the goal for extensive government interventions in foodgrains markets. The policies were aimed at preventing private traders from undercutting government policies on procurement, storage, and distribution. Specifically, as a first priority, the government required farmers and millers to sell adequate quantities to government agencies to meet their supply targets. Stocking limits, zoning laws, and intracountry movement restrictions were also imposed on private traders. Bangladesh, however, put into abeyance the 50-year-old anti-hoarding act that had imposed stocking limits on private traders. Certain controls on intracountry grain movement in India have been more or less removed while others, such as compulsory procurement of rice from millers, continue to exist. In Pakistan, although the government monopoly on rice procurement and distribution was phased out and private trade was allowed, large-scale interventions and controls continue to prevail in wheat, the country's main staple.

International Trade Policies in South Asia

International trade restrictions took several forms, from banning private-sector trade and requiring trade licenses to increasing tariff and imposing quantitative restrictions. With the overall goal of helping farmers, import controls were meant to stop competition while export controls were maintained to ensure high availability and low prices domestically. These policies often had just the opposite effect, however, and farmers in the region were left completely unprotected when they received less for their product than the prevailing international price.

Sri Lanka Liberalized its trade policy in 1977. In 1993, the country gave up its monopoly on rice imports and no longer required licenses on imports by 1996. In 1988, Nepal adopted a broad structural adjustment program that required elimination of all quantitative restrictions. In 1993, Bangladesh followed, with the legalization of private international trade in wheat and rice at modest tariff rates. Both India and Pakistan have allowed private trade in grains, removed several export controls and licensing requirements, and replaced quantitative import restrictions with tariff. However, in 2007–08, India restored export controls in response to high international food prices.

Cost-effectiveness of Public Intervention in Domestic Grain Markets

In India and Pakistan, the magnitude of government intervention in domestic grain markets continues to be very large and fraught with serious inefficiencies. The marketing cost of private traders is only about 70 percent of the cost of the Food Corporation of India. Similarly, the "incidental" costs for Pakistan's Agricultural Storage and Supplies Corporations were 15–27 percent higher than for private traders in 2000–2003; in Nepal from the 1980s until the early 2000s, marketing costs for private traders were about 41 percent lower than for the Nepal Food Corporation. Bangladesh—where private traders have a slightly higher (by about 1 percent) cost than the government—seems to be the only exception.

Recently, the basic criticism of government involvement in procurement, storage, and distribution is that it results in significant price distortions. In Pakistan, the real prices of commodities remained more stable than the prices of commodities that involved government intervention. The Indian experience in intervening in grain markets highlights the problems with public procurement, stocking, and distribution at administratively fixed prices. If it aims to materially benefit the consumer, such a system will either impose an implicit tax on farmers or require enormous subsidies, which may not be fiscally sustainable.

Impact of Border Trade Reforms

The country case studies in Liberalizing Foodgrains Markets clearly show that in net importing countries such as Bangladesh, Nepal, and Sri Lanka (and, until recently, Pakistan), trade liberalization will increase import competition, which will lower domestic prices. Lower prices benefit consumers but harm producers. Liberalization has also helped smooth seasonal fluctuations with the narrowing of the seasonal highs and lows in rice prices between the 1980s and 2000s.

Since the mid-1990s, India has replaced Thailand as the main supplier of rice into Bangladesh, and the real price of rice has fallen in Bangladesh by 2.57 percent per year from 1981 to 2003. That reduction in rice prices clearly benefited poor consumers, as seen in the narrowing of the gap in per capita consumption between the bottom 40 percent and the top 20 percent of the population. Similarly in Nepal, nearly 10 percent of rice consumed comes from India, which has benefited 60 percent of Nepal's landless households.

Following trade liberalization in Sri Lanka, rice prices have fallen by 34 percent, resulting in a 45 percent rise in demand and a 33 percent rise in calorie intake. However, the adverse incentive effect on farmers resulted in a fall in rice supply by 28 percent. Pakistan has recently turned into a net exporter of its main staple, wheat, which remains relatively untouched by the trade reforms there. Similarly, as a net exporter of cereals, trade liberalization in India results in higher domestic prices that benefit farmers but hurt consumers.

Spatial Integration of Domestic Markets

Spatially integrated markets ensure that local shortages in certain food items do not translate into a sharp rise in food prices. Removing the restrictions and constraints on intra-country movement of goods can result in large efficiency gains that can be reaped fairly quickly. Difficult physical terrain—such as hills and mountains in Nepal and extensive rivers in Bangladesh—requires sustained investment in transport infrastructure to achieve greater spatial integration. Grain markets in Bangladesh that invested in roads and river-transport infrastructure are better spatially integrated than those in Nepal, where lack of investment in transport infrastructure often results in local scarcities in remote hills and mountains as well as often significant price differentials.

The case studies in India and Pakistan demonstrate that government policies that regulate the movement of goods across states also result in a lack of spatial integration. The findings for India showed that large-scale gains in efficiency arose with the removal of zoning restrictions because then private trade could occur between neighbouring states driven by arbitrage possibilities rather than their deficit–surplus status. This is in contrast to the practice of government agencies, which traded directly between surplus and deficit states over long distances. Additionally, the removal of zoning laws combined with external trade liberalization allowed some deficit states to source their food from international markets rather than from surplus states within the country, because of differences in the international and domestic transport costs.

Role of Domestic Market Structures

In agricultural markets, especially food markets, traders have relatively more market power than either farmers or consumers. This works in two ways: Traders have oligopsony—that is, the market position of many sellers with few buyers—in their transactions with farmers while having the position of oligopoly—many buyers with few sellers—in their transactions with consumers. This strong position allows them to be price fixers and not price takers in both transactions. The large margins for traders are often touted as the justification for government intervention in agricultural markets.

The Sri Lankan case studies suggest that losses to paddy producers from trade liberalization can certainly be minimized if the oligopsony power can be restrained and border reforms implemented. In Bangladesh, absence of large-scale government intervention in the output market is a major cause for the increase in competition in trading services. The number of rice traders increased twelvefold in approximately 30 years—from 4,000 in the 1960s to more than 48,000 in the 1990s. This rise in competition between traders has ensured that farmers are able to (1) avoid distress selling, (2) hold a fairly high proportion of on-farm stocks, and (3) sell their produce at the farmgate rather than in distant markets, thereby avoiding transportation costs.

Country Policy Changes

Divergence in the evolution of South Asian food policy measures over the past four decades indicates that each country in the region faces different challenges. In order to potentially achieve higher gains from agricultural production and stronger markets from better promotion, the Government of Nepal must invest in infrastructure and the Government of Bangladesh in irrigation, storage, transport, and communication. Subsidizing general purpose public goods such as agricultural research and extension services will benefit all farmers—small and large—while input subsidies benefit only a few farmers who use those inputs.

In Sri Lanka, maintaining stable trade policies—rather than varying the tariff rates frequently imposed in response to domestic supply conditions—will be a more effective way to achieve relatively steady levels of domestic consumption. These and effective measures to improve agricultural productivity will help Sri Lanka attain food security sooner than it would under the current set of policies. As food-surplus countries, India and Pakistan must undertake further measures to correct market imperfections.

It is time to restrict the government control over private participation in trade and maintain orderly conditions in the foodgrains markets.

Promoting competition, devoid of policy-induced distortions, will help improve resource allocation by farmers, as well as allow consumers to benefit from competitive prices of food. In a competitive environment, the private sector can also deliver in other spheres with a direct bearing on agriculture and food security, such as in agricultural extension services, value addition in agriculture, and the development of infrastructure.

Intraregional Trade

There has been a rise in intraregional trade in rice and wheat within South Asia. Nepal and Bangladesh, unlike Pakistan and Sri Lanka, have relatively porous borders with India, and, therefore, would feel the impact of input- and output-subsidy policies in India (as, perhaps to a lesser extent, would Sri Lanka). This clearly emphasizes the need for greater policy coordination between the countries of the region and especially greater policy discipline on the part of India. The platform of the South Asian Association for Regional Co-operation should be explored to further strengthen policy coordination.

The food price crisis brought out the stark consequences of uncoordinated actions aimed at achieving greater national food security. In November 2008, recognizing that surplus and deficit countries are neighbours, members of the South Asian Association for Regional Co-operation proposed to set up a regional granary to which all countries would contribute and which, in turn, could be used to help any member country tide over idiosyncratic supply shocks. The experiences of the food stocking, distribution, and domestic trade policies of individual South Asian countries suggest that a regional food bank would prove largely ineffective unless trade within the region was also liberalized—just as internal movement restrictions within a country impede government efforts to smooth prices. Hence, free trade agreements, such as those provided for in the South Asian Free Trade Area, complement effective food security through collective action. Moreover, the realities in South Asia also suggest that the design of such regional food reserves would be complicated less by economics than by the complex politics of the region.There is potential for cooperation between South Asian countries in the transport of food commodities and the direct related implications for food security. For example, it could be advantageous for certain states in India to transport goods to northeastern Indian states by shipping them through Bangladesh.

Effective food policy measures have great potential to reduce poverty and, hence, enhance economic growth. It is the combined effect of evidence-based policy reforms, good governance, and liberalized markets that will ensure food security for the current and future generations of South Asia.

Emerging Issues

Market liberalisation, including the reform of agricultural and food markets, is a key element of economic structural adjustment programmes. The origins of the term structural adjustment rest with the entry of the World Bank into policy-based lending, in the early 1980s. It refers to restructuring production so that the allocation of resources is based upon opportunity cost and in this way optimises the outputs of those resources. Thus structural adjustment is intended to lead to economic growth. Originally the supply side policies of the World Bank were pursued independently of the economic stabilisation programmes of the IMF, which deal more with the demand side of an economy. However, since the mid-1980s the two organisations have worked together to ensure that structural adjustment and stabilisation programmes have complemented one another.

Before the benefits of ESAP can be realised, policy makers must arrive at an agenda of policy and institutional reforms and design a sequence of measures which is technically consistent and politically sustainable. Unfortunately, there is little informed guidance available to them in their task, because there is no substantive empirical understanding of best practice in structural adjustment. It is likely to be a considerable time yet before enough data has been collected and analysed to provide a basis for guiding policy makers.

A major question for most LDCs that seek to implement ESAPs is what role, if any, to assign to agricultural marketing parastatals after markets have been liberalised. The options range from disbanding them altogether to privatising either the organisations themselves or their functions. These are not easy decisions. If these parastatales are abolished then there has to be solid evidence that there are other organisations both able and willing to perform their functions. Some of these functions may prove wholly unattractive to commercial enterprises but must be performed by one agency or another (e.g. management of the strategic food reserve and developmental projects). If, alternatively, the decisions is to privatise an agricultural marketing parastatal then, in most cases, difficulties such as accumulated deficits, credibility with lending institutions and future tax liabilities first need to be overcome. More often than not encouraging the private sector to invest in liberalised markets is an integral component of ESAPs. It is reasoned that a competitive private sector can greatly improve the efficiency of a marketing system. There are, however, a number of impediments to the participation of the private sector when markets are opened up to competition for the first time. These include a lack of trust, on the part of prospective entrepreneurs, that market liberalisation measures will not be reversed, trade licensing and regulatory arrangements that are perceived to be too restrictive, the continuance of any form of price control and uncertainty that surviving marketing parastatales will not be given unfair advantages over them. Other common impediments are inadequacies in marketing infrastructure, access to credit and access to marketing information. Private sector participation in liberalised agricultural markets is also influenced by macro-economic as well as micro-economic policies. Exchange rate policy, economic stabilisation measures and the inflow of economic aid to the country all influence the incentive to private entrepreneurs to enter the liberalised market.

Governments that sincerely wish to encourage private sector participation in the marketing system should ensure that policy statements are widely communicated and unambiguous in the intent to promote private sector involvement; should develop a supportive regulatory framework; need to provide technical and entrepreneurial training; have to initiate improvements in the marketing infrastructure; promote the establishment of a marketing information system to guarantee market transparency and help prospective participants to gain access to formal sources of credit.

References

Scarborough, V. and Kydd, J. (1992), *Economic Analysis of Agricultural Markets:* A Manual. Chatham, UK, Natural Resource Institute, p. 140.

Smith, L. (1992), *Agricultural Sector Performance and Policy And the National Economy.* Network And Centre For Agricultural Marketing Training In Eastern And Southern Africa, p. 74–80.

Spooner, N. and Smith, L. (1989), *Structural Adjustment Policy Sequencing in Sub-Saharan Africa,* Rome: Economic and Social Policy Division, Food and Agriculture Organization.

3

Storage of Foodgrains

The storage of foodgrains has been an age long practice with cultivators and traders. More pest-free storage is needed for handling crops at harvest time and to carry over reserves from year to year. Considerable losses both in quality and quantity of food-grains take place in storage due to a number of factors.

Organisms directly responsible for causing loss in stored products are insects, mites, rodents, fungi and bacteria. Among them, insects and mites are the most important hazards to the safe storage of grains. The insects that attack stored grains are rather general feeders, but some of them prefer certain grains. It is estimated that 5-10 per cent of the stored grain is lost every year due to insect damage in India. Some examples of the major insect pests of stored grain are *Rhyzopertha dominica* (lesser grain borer), *Oryzaephilus surinamensis* (saw toothed grain beetle), *Sitophilus oryzae* (rice weevil), *Trogoderma granaria* (khapra beetle), *Tribolium castaneum* (red flour beetle) and *Plodia interpunctella* (Indian meal moth).

Whether storage takes place on the farm or in silos off the farm, increases in the value of products due to their time utility must be sufficient to compensate for costs at this stage, or else storage will not be profitable. These costs will include heating, lighting, chemical treatments, store management and labour, capital investment in storage and handling equipment, interest charges and opportunity costs relating to the capital tied up in stocks. Among the less tangible costs is the risks attached to storage.

These include shrinkage due to pilferage, pests, fungal growths and loss of quality due to ageing. Another risk is that demand could fall with adverse effects on prices.

Since the advent of structural adjustment programmes and market liberalisation, some grain marketing parastatals have lost their monopoly of the market and consequently the volumes of grains which they are handling has dropped substantially. This means that they no longer require all of their storage capacity and a number of marketing parastatals now rent some of their storage capacity to farmers, grain traders and other participants in the grain marketing system.

Two types of storage facility, are commonly found, namely: the bulk storage facility where cereals are stored in concrete and/or metal bins, and the bag storage facility where the crop is stored either inside a warehouse or in the open and then covered by tarpaulin sheets. In comparative terms, the advantage of a bulk over a bag storage system are that it is more efficient because it:

- reduces congestion at the depots by not allowing for bagged maize to be dumped all over the depot yard
- reduces handling costs
- saves foreign exchange on bags, tarpaulin and fumigation, and lowers storage losses.

Its disadvantages are that:

- the initial invest is high, with a significant foreign currency component
- it is inflexible in terms of not being easily expandable to cope with changes in intake and off-take levels
- it relies heavily on an efficient transport system because a silo complex is only economically viable when throughout is at least 1.6 times its capacity, and
- it needs skilled manpower to run and maintain the entire system.

On the other hand, a bag system ideally overcomes the problems associated with a bulk system as enumerated above. Its main disadvantages are:

- higher quality losses in storage due to insect pests and rodents
- higher demand for foreign exchange associated with the purchase of bags
- tarpaulin and fumigation sheets.

The depot manager controls the day-to-day operations of the depot. The duties of the depot manager include accepting, checking the quality, recording the quantity and storing the produce brought in by the farmers. Each depot's record of accepted and stored produce are then sent to headquarters for necessary action. With this information to hand, inter-depot or inter-district transfers can be effected.

Importance of Storage

Good storage facilities are important to the farmers all over the world. They help to ensure household and community food security until the next harvest and commodities for sale can be held back so farmers can avoid being forced to sell at low prices in the glut that often follows a harvest. Though considerable losses occur in the field, both before and during harvest, the greatest losses are noticed during storage.

1. Loss in Quantity

Losses of the food grains in terms of weight are quantitative losses. Insects, rodents, birds etc. feeds on the product causing weight loss. These weight losses are not always apparent. For example, some insects eat only the centres of grain kernels so, even though the volume of grain may appear to remain the same, there can be considerable weight loss.

2. Loss in Quality

Losses of this type can be nutritional, chemical, through contamination with toxic moulds or foreign matter. Pests that selectively eat a part of the food-stuff (such as the nutritious germ of the grain) will reduce the value of the food-stuff as a whole. Also, there is the loss of vitamins through the action of sunlight and temperature. Chemical changes are particularly common in fatty foods through the development of rancidity.

Aflatoxin (a toxic substance) producing moulds like *Aspergillus niger*, can grow on many products which pose a long term health risk. As maize, coconut and peanuts are particularly susceptible, storage of these products need special care.

General contamination can result in many ways and shows up in the form of insect fragments, rodent hairs, excreta and urine, as well as dust and other materials that enter the product through human mis-handling. The presence of rat urine can cause serious problems as rats are carriers of Weils disease. Sieving is often used to reduce the obvious signs of foreign matter contamination.

Farmers store the produce for two reasons; for home consumption and marketing. Farmers may not accept improvements which incur costs when storing primarily for home consumption. In many cases, these need only small improvements to make the difference between simply having enough for subsistence and creating a surplus for sale.

In most countries grains are among the most important staple foods. However they are produced on a seasonal basis, and in many places there is only one harvest a year, which itself may be subject to failure. This means that in order to feed the world's population, most of the global production of maize, wheat, rice, sorghum and millet must be held in storage for periods varying from one month up to more than a year. Grain storage therefore occupies a vital place in the economies of developed and developing countries alike.

The market for food grains is characterized by fairly stable demand throughout the year, and widely fluctuating supply. Generally speaking people's consumption of basic foods such as grains does not vary greatly from one season to another or from year to year. The demand for grain is 'inelastic', which means that large changes in the market price lead to relatively small changes in the amount of grains which people purchase.

Market supply, on the other hand, depends on the harvest of grains which is concentrated within a few months of the year in any one area, and can fluctuate widely from one year to the next depending on climatic conditions. New varieties that have shorter growing periods, and variation in climatic conditions and farming systems in different regions of a country, can help to even out the fluctuations in market supply. But even in a country such as Indonesia, which has diverse climatic and farming conditions and where 90 per cent of rice land is under short duration high yielding varieties, about 60 per cent of production is harvested within a three month period.

The main function of storage in the economy is to even out fluctuations in market supply, both from one season to the next and from one year to the next, by taking produce off the market in surplus seasons, and releasing it back onto the market in lean seasons. This in turn smooths out out fluctuations in market prices. The desire to stabilise prices of basic foods is one of the major reasons why governments try to influence the amount of storage occurring, and often undertake storage themselves.

Who Stores and Why?

Farmers, traders and governments all have reasons for storage other than the profitability of the storage enterprise itself. Storage is a component within a farming system, a trading enterprise, or a government policy, and may be undertaken because of its contribution to other activities or objectives within these broader contexts.

Farm Storage

For small farmers the main purpose in storing grains is to ensure household food supplies. Farm storage also provides a form of saving, to cover future cash need through sale, or for barter exchange or gift-giving. Grain is also stored for seed and as inputs into household enterprises such as beer brewing, or the preparation of cooked food. When there are significant inter-seasonal price variations, small farmers often store for speculative gain, that is to say they 'play the market'. This is most common in more prosperous areas, such as the Southern Highlands of Tanzania and southern Mali, which produce a mixture of cash and food crops, and where farmers' financial circumstance make it easier for them to sell when the price is best. Speculative considerations are even more important in the storage decisions of large-scale commercial farmers.

Despite the desire to store grain in order to cover food requirements and future cash needs, farmers often sell a large proportion of their produce at harvest, when prices are low. This is frequently the case with deficit producers, who must satisfy cash needs immediately after the harvest, only to buy in food later in the season.

There is an ongoing debate about whether farmers are forced to sell because of debt and economic dependence on others, or whether they sell because they regard storage as too costly (in terms of time), or too risky (given the risk of losses and unpredictability of future prices), or unprofitable in relation to other investments such as cattle. There is no single answer to the debate, since there is much variation in the circumstances under which individual farmers operate, both within and between nations. The 'forced sales' situation has been documented by some authors in South Asia, while the 'wise farmer' been found to apply in South-East Asia by Mears and Ellis et al. In the Sahelian countries of Africa, conflicting findings have been reported. Carefully documented work by Dione has shown head taxes in Mali to have resulted in forced sales, but Berg and Kent report several authors who have reached opposite conclusions.

Another reason for not storing is the unpredictability of future prices, which often makes storage a risky business. This is particularly the case in countries such as Mali where prices vary widely from year to year, and do not follow a steady monthly trend. Malian prices normally rise between harvest time and the lean season, but farmers who engaged in speculative storage in 1989 suffered significant losses. Storage was particularly risky as Mali was passing from a period of scarcity, when prices levels were related to the cost of imports, to a period of surplus, when they were related to the price at which Mali could export. Movement between surplus and deficit in the Sahelian countries probably explains the wide variation in Mali's prices from year to year.

Farmers may sell their grain at any time from maturity (sale of the standing crop) onwards. Sale at or before harvest has the advantage that the farmer is saved the cost and time involved in preparing the crop for storage. Transport, threshing, winnowing and drying are all passed on to other levels in the marketing chain, leaving the farmer free to attend to the next crop, or to other farm or off-farm activities. It has been estimated that post-harvest activities account for one quarter of the total cost of production even for small farmers in poor countries. Early sale also reduces the risks of losses in postharvest activities, and this is particularly advantageous in cases where the harvest occurs in the wet season.

Trader Storage

The role of traders in cereal storage varies enormously between different parts of the world and between different crops. In most African countries traders carry out very little interseasonal storage of coarse grains, but buy and sell quickly, earning a moderate profit on each transaction. Most storage is carried out by farmers, and to a lesser degree by Government marketing boards and consumers who buy in anticipation of future household needs. Given a general situation of capital shortage, long-term storage of staple grains is insufficiently profitable to attract the interest of traders, who can earn more money by investing in fast moving consumer goods.

However the opposite is often true with rice in Africa. This crop is generally produced as a cash crop for the urban markets, but does not have a major demand for use as a staple in rural areas. Often much of the rice is imported and this has encouraged the emergence of large traders able to obtain finance major shipments and to negotiate advantageously with the authorities. Even when sourcing supplies from local producers, traders and millers must hold stocks to cover the needs of their urban clientele, and cannot rely on steady supplies arriving from rural areas.

In Asian countries, traders have a much larger role in interseasonal storage. The two major cereals are rice and wheat and both of these must be milled before reaching the consumer. This is unlike the situation in most of Africa where coarse grains such as maize, millet and sorghum are the main staples. Typically African consumers buy these grains whole, and either grind them at home or take them to be ground at small custom-mills.

Large millers who become involved in the marketing chain tend to have good banking connections and can obtain capital at reasonable cost. Studies by the Natural Resources Institute (NRI) in Indonesia and Pakistan indicate, that wherever Government policy is conducive, millers enter the storage business on a large scale. In Indonesia, traders and millers store about 50% of that part of the rice crop which is carried over from the first harvest. Indeed it is common for them to store beyond the point when storage is profitable in its own right. This is because storage is only part of a business activity which involves milling and distribution of milled rice; millers must store in order to keep the mills running out of season, and to maintain supplies to regular customers. Losses on storage are more than compensated for by the gains on other operations.

In Pakistan, the millers' role in wheat storage has been limited by Government subsidies to public sector institutions, which procure about 60% of the marketed portion of the wheat crop. Rather than procuring wheat themselves, millers found it cheaper to procure from these Government institutions which carried out most of the long-term storage. However, when the Government raised their selling price in 1989 and thereby improved the incentives for millers to store, these responded promptly by buying up more stock.

In the future, storage behaviour in African countries will probably evolve towards the Asian pattern. The liberalisation of cereal markets will encourage the development of the private trade, the reform of banking systems should gradually increase traders' access to capital markets, and increased urbanisation and sophistication of tastes will favour the emergence of large milling enterprises.

Government Storage

Government may become involved in storage for the purpose of stabilising prices and revenues to farmers. Related to this is Governments' overriding concern for national food security, which is fundamental to political stability. Governments therefore use storage to balance national supply and demand over time, and to minimise the risk of politically embarrassing shortages. They are thus attempting to supplement, and in some cases to replace, market mechanisms, on the assumption that the market can only achieve the balance with an unacceptable degree of supply and price fluctuation.

Governments do not involve themselves in the grain market only for reasons of national interest: they are often concerned with rewarding or placating particular lobbies or sectional interests. In developed countries farmers' interests often receive a high priority in Government decisions, out of proportion to their numbers. High 'support prices' encourage production in excess of demand, and surpluses have to be stockpiled at the taxpayers' expense. In many developing countries, the interests of the civil service and ruling party often take priority. Large national food reserves tend to be supported by the civil servants whose job it is to manage them, and by politicians who sometimes use their procurement and distribution as a means of dispensing patronage.

Governments may keep different types of storage reserve, depending on how much they wish to intervene in the grain market. Some of the options are:

a. a food security reserve to be sold or distributed for free at times of extraordinary shortage or famine. Such reserves can be found in Sahelian countries like Mali and Chad. They are of limited size (e.g. 10% of the normal volume of grain marketed crop), and are usually limited to the amount thought necessary to tide the country over until the arrival of food aid or imports. They are not designed for the purpose of stabilising prices to producers and consumers.

b. a price stabilisation stock, as in the case of Indonesia. Here the Government has no monopoly role in grain procurement and distribution but buys and sells grains in competition with private operators. Average interseasonal retail price increases in Java are only 11% of the lowest monthly figure. How much this extraordinary low figure is due to Government stockholding and how much to the stockholding activities of millers is a matter of debate.

c. national storage reserves designed to supply most or all consumer needs in urban areas, and in rural deficit areas. In this case the Government has either a statutory trading monopoly, or a monopoly of all interregional shipments, and is the only party allowed to store significant quantities of grain. Between the 1960s and the early 1980s, such systems were the norm in many African countries, before the onset of liberalisation. Even now, the grain marketing systems in some countries, including Zimbabwe and Kenya, are still partially structured in this way.

Such Government operations usually benefit from public subsidy (intended or de facto) and capital investments are largely financed by overseas aid. Indeed subsidies are necessary if the public sector is to carry out functions which would not be profitable to the private sector. However, in some countries subsidies have allowed the State to 'crowd out' private sector competition. In the case of Pakistan for example, this phenomenon has resulted in the State handling about 60% of the marketable surplus. In many countries, such competitive advantages are outweighed by the high cost of fulfilling Government requirements (e.g. to buy and sell at fixed politically-determined prices, and to supply civil servants' consumption needs), overstaffing and slow decision-making processes. In such countries e.g. Tanzania in the 1980s, the official marketing agency may become insolvent and be gradually displaced by private sector competition.

Even relatively efficient Government trading operations face the problem that the more grain they buy, and the more they succeed in stabilising prices throughout the year, the less the incentive for private sector storage'. The responsibility for storage then falls very heavily on Government, and the private traders and millers concentrate on buying and selling quickly. Consequently, the Government finds that it has very high storage costs which it cannot recover through sale prices which have been politically determined. In the end the Treasury or Government banks must bale out the Government enterprise, thereby increasing the budgetary deficit.

Since 1981, there has been a major move to liberalise grain marketing systems in developing countries, and this has been stimulated by both donor pressure and the massive budgetary deficits stemming from the operation of Government marketing boards. Many African Governments are opting for the first of the above options i.e. a limited food security reserve.

Some countries do not appear to need any reserve stocks but can rely on international trade to assure food security and to stabilise prices. This is particularly the case with some deficit countries in Africa, such as Swaziland and Namibia, who have good communications with the world market and are close to major grain suppliers.

Lastly there are some countries where it would seem most appropriate for Government to maintain some sort of price stabilisation role. Such is the case in landlocked countries like Zimbabwe and Malawi, whose production fluctuates between surplus and deficit. If the Governments of these countries totally withdraw from price stabilisation, prices are likely to be subject to very wide interannual fluctuations, with adverse effects on production incentives and consumer welfare. Nevertheless these countries still have major scope for liberalisation. By improving port facilities and communications with the outside world, and by developing intra-regional trade, they can greatly reduce the required level of stockholding.

The move towards liberalisation in developing countries contrasts with the situation of the developed countries, where Governments are still heavily involved in the grain trade. Developing country officials often ask why they should be asked to liberalise while rich countries fail to do so. The answer is simply that these countries have the wealth to support their farmers at the expense of their non-farming majorities.

Farmers, traders and governments all have reasons to store grain, but they also have reasons for limiting the amount of storage. The unit costs of storage tend to be constant (or to decrease slowly) as larger quantities are stored, but the benefits fall off as more is stored. In deciding how much to store, the benefits must be balanced with the costs involved.

Principles of Storage

Most developing countries are in the tropics, often in areas of high rainfall and humidity. These conditions are ideal for the development of micro-organisms and insects which cause high levels of deterioration of crops in store. Food losses during storage are the result of biological, chemical or physical damage.

In order to reduce the amount of food grains lost, the environment in the store needs to be controlled so as to lower the possibility of:

- biological damage by insects, rodents and micro-orgamsms.

- chemical damage through rancidity development and flavour changes, etc.
- physical damage through crushing, breaking, etc.

Good storage thus involves controlling the factors, like temperature, moisture, light, pests and hygiene.

I. Temperature

The temperature within a store is affected by the sun, the cooling effect of radiation from the store, outside air temperature, heat generated by the respiration of both the food in store and any insect pest present.

Most of the micro-organisms thrive between 10 and 60°C temparature where as insects between 16 and 45° centigrade. Normally, in tropics and sub-tropics storage temperature lies between 25 and 35°C which is favourable for the survival of the micro-organisms and insects. Direct temperature control is not usually possible, so other measures, particularly reducing the moisture content of the stored produce, are necessary.

Running is a method of controlling insect pests involving heat. The produce will be laid out in a thin layer in the hot sunlight. At high temparatures (40 to 45° C) the insect pests tend to leave the grain. It should be noted, however, that running does not always destroy eggs or larvae.

Improper maintenance of storage temperature can result in biological and chemical damage to the food stuff being stored. Examples include the loss of germination ability in seed materials and the accumulation of sugars in some commodities which need relatively low storage temperatures.

Temperature also controls chemical damage. The speed of chemical change in a food depends upon the temperature and the food's moisture content. A 10°C rise in temperature causes an approximately two-fold increase in the rate of reaction. Thus, cold storage will retard such changes as fat oxidation and vitamin loss. Many dried food grains benefit from even a small reduction in their storage temperature, and cool and dry conditions can greatly reduce the rate of development of brown discolouration and off-flavours. Physical damage involves melting of fats in the products at high temparatures and crystallization of sugars in sweet foods at low temparatures.

2. Moisture

All micro-organisms, including moulds, require moisture to survive and multiply. If the moisture content in a product that is to be stored is low, micro-organisms will be unable to grow, provided that the moisture inside the storage structure is also kept low. Moisture should therefore, be prevented from entering the store.

All materials that have been dried will try to come back into equilibrium with the climate around them. In tropical countries this usually means absorbing moisture. The moisture level below which micro-organisms cannot grow is referred to as the safe moisture content. The Table 1 below lists the safe moisture content levels for cereals and pulses valid to temperatures up to 27° centigrade. Slight variations in safe moisture contents arise depending upon the variety.

Table 1. Safe moisture content levels

Sl.No	*Safe moisture product*	*Content (%)*
1	***Cereals: Maize (shelled)***	13.5
	Maize flour	11.5
	Paddy rice	15.0
	Milled rice	13.0
	Sorghum	13.5
	Millet	16.0
	Wheat	13.5
	Wheat flour	12.0
2	***Pulses: Broad bean,cow pea***	15.0
	Lentil, pea	14.0

While in general it is essential that all food-stuffs are below their safe moisture content before they enter the store, the safe moisture content is to some extent related to the required storage time. Moisture levels above the safe moisture context can be tolerated if only short storage times are required.

Condensation of moisture can cause storage problems. If the walls of a store are cooled below their dew point by low night temperature, condensation can occur and increase the moisture content in the layers of the produce. The siting and the ventilation of the store is important.

It is also important to note here that most stored food products are "alive" and respiring, thus giving off moisture, as well as heat.

Tips for Storage of Food Grains

This has been reported that up to 10 million tonnes of food-grains are lost each year in India through faulty storage techniques. If saved, it would have been more than enough to meet any world food deficit. It has been estimated that 10 percent loss in storage may be sufficient enough to feed 72 million people. The large stocks of grain in storage pose various problems which are unique and quite different from those of smaller lots. Therefore, adequate attention must be paid in the large scale storage of food-grains. To minimize the storage losses some practices for large scale storage of food grains has been developed.

1. The grains must be checked in the field before harvest to make sure that the grains are free from insects and diseases.
2. The harvesting and transporting implement must be cleaned before a new crop is harvested.
3. The grains must be harvested and threshed to avoid any breakage of grains, because broken grain will not store well.
4. The threshing yards which are free from insect infestation should be used.
5. Before storage the grains should be cleaned and graded. Unclean grain contains small amounts of straw, weed seeds and dirt, which not only decrease the value of food-grains but also cause the grains to deteriorate during storage.
6. No food-grains with a moisture content higher than the safe acceptable level should be accepted for storage. Dry the moist grain before storage because, it respires more quickly and gives off more heat and moisture, which encourages build of insect population and mould growth and hot spots develop in bulk grains.
7. The grains should be spread over plastic sheets or cemented floor while drying, otherwise it will pick up the moisture from the ground. The grains should be kept cool and dry between the time of harvest and storage.

Grain Drying

There is an essential need to dry grain quickly and effectively after harvest and before storage to retain maximum quality, to attain a moisture content sufficiently low to minimise infestation by insects and microorganisms (bacteria, fungi, etc.), and to prevent germination.

Wherever possible, it is traditional to harvest most grain crops during a dry period or season and simple drying methods such as sun drying are adequate. However, maturity of the crop does not always coincide with a suitably dry period. Furthermore, the introduction of high-yielding varieties, irrigation, and improved farming practices has led to the need for alternative drying practices to cope with the increased production, and grain harvested during the wet season as a result of multi-cropping.

Natural methods of drying make use of exposure of the wet grain to the sun and wind. Artificial dryers employ the application of heat from combustion of fossil fuels and biomass resources, directly or indirectly, and in both natural and forced convection systems. Mechanical dryers, long used in developed countries, are finding increased application as farming and grain handling systems develop.

In the process of drying heat is necessary to evaporate moisture from the grain and a flow of air is needed to carry away the evaporated moisture. There are two basic mechanisms involved in the drying process; the migration of moisture from the interior of an individual grain to the surface, and the evaporation of moisture from the surface to the surrounding air. The rate of drying is determined by the moisture content and the temperature of the grain and the temperature, the (relative) humidity and the velocity of the air in contact with the grain.

Natural and Solar Drying

Sun Drying

The traditional practice of grain drying is to spread crop on the ground, thus exposing it to the effects of sun, wind and rain. The logic of this is inescapable; the sun supplies an appreciable and inexhaustible source of heat to evaporate moisture from the grain, and the velocity of the wind to remove the evaporated moisture is, in many locations, at least the equivalent of the airflow produced in a mechanical dryer. In tropical countries, for at least several months of the year, the mean level of insolation upon the ground is more than 0.5 kW/m^2 (measured as a mean over the hours of daylight). The heat available therefore, assuming a 12 hour day, is 21.6 MJ/m^2, a quantity theoretically sufficient to evaporate 9 kg of water.

Even today, sun drying of grain remains the most common drying method in tropical developing countries. It is first employed when the crop is standing in the field prior to harvest; maize cobs may be left on the standing plant for several weeks after attaining maturity. Although not requiring labour or other inputs field drying may render the grain subject to insect infestation and mould growth, prevent the land being prepared for the next crop and is vulnerable to theft and damage from animals. Drying in the field may also be carried out after harvest with the harvested plants laid in stacks with the grain, maize cobs or panicles raised above the ground and exposed directly to the sun. Data on the drying of paddy in the field has been gathered by Angledette and Mendoza et al.

Drying on flat exposed surfaces is the most common way of drying grain after harvesting and threshing. For drying small amounts on the farm grain may be spread on any convenient area of land. Contamination with dirt cannot be easily avoided with this method and cleaner dried grain can be obtained by drying the grain on plastic sheets, preferably black.

Purpose-constructed drying floors are commonly used where there is a need to dry large quantities of grain during the season, e.g. at most rice mills. The floors are usually made of concrete or brick, these materials presenting a relatively smooth and hardwearing surface. Floors should be constructed to withstand the movement of vehicles and sloped or channelled to hasten the runoff of rainwater. The paddy is spread in a thin layer on the floors and raked at intervals, preferably 7-8 times daily, to facilitate even drying. At night the paddy is heaped into rows and covered with sheeting.

Work by Chancellor and Soetoyo & Soemardi has demonstrated that paddy can be dried from 24-26% moisture to 14% moisture at depths of 50-100 mm at a rate of 3.3 kg/m^2.h for stirred paddy and 1.9 kg/m^2.h for unstirred paddy. The grain can reach temperatures as high as 60°C under clear skies and the rate of drying can be extremely high. Under these circumstances kernel cracking and loss of head rice can be appreciable, particularly if paddy is dried to below 14% moisture. Covering the paddy around midday may be beneficial under particularly hot and sunny conditions. Experiments at IRRI have shown that cracking can be reduced by 25% if paddy is dried in the shade but the benefit from the improved quality is generally more than offset by the longer drying times and hence reduced throughput and increased costs.

In rainy weather, even though drying will be slow, every effort should be made to prevent wet freshly-harvested paddy from over-heating with deterioration in quality by spreading on floors rather than let it remain in heaps and sacks. Under these conditions or when there is great demand for drying space paddy can be dried to 17-18% moisture and then temporarily stored for 15-30 days before final drying.

Crib Drying

Compared with paddy, cob maize can remain at relatively high moisture contents, in excess of 20% with natural ventilation for considerably longer periods, from one to three months. The maize crib in its many forms acts as both a dryer and a storage structure. The rate and uniformity of drying are controlled by the relative humidity of the air and the ease with which air can pass through the bed of cobs. The degree of movement of air through the loaded crib is largely attributable to the width of the crib; research in West Africa has shown that crib widths should not exceed 0.6 m (Anon 1980). Guide-lines on crib design, construction and operation have been prepared by Bodholt

Solar Dryers

An improved technology in utilizing solar energy for drying grain is the use of solar dryers where the air is heated in a solar collector and then passed through beds of grain. There are two basic types of solar dryer appropriate for use with grain: natural convection dryers where the air flow is induced by thermal gradients; and forced convection dryers wherein air is forced through a solar collector and the grain bed by a fan.

The solar chimney provides a column of warm air that increases the thermal draught of air through the dryer. It is made of a bamboo frame covered with black plastic sheet. In Thailand paddy was dried from 20% moisture to 13% moisture in 1-2 days and the rice quality was appreciably greater than that from sun dried paddy. A disadvantage of the dryer is its high structural profile which poses stability problems in windy conditions, and the need to replace the plastic sheet every 1-2 years. A smaller (100 kg capacity) and simpler version of this type of dryer has also been developed.

The forced convection solar dryer can be considered as a conventional mechanical drying system in which air is forced through a bed of grain but the air is heated by a flat plate solar collector rather than by more conventional means.

The performance of a flat plate collector can be quantified by calculation of the collection efficiency; the ratio of the heat gathered by the collector to the insolation incident on its surface. The collection efficiency is a function of the air velocity through the collector, the geometry of the air duct, the absorptivity of the absorption surface, and the transmissivity of the cover(s).

Considerable work has been undertaken in developing low-cost and efficient solar collectors for crop drying applications. The simplest type of collector is the bare plate which consists simply of an air duct the uppermost surface of which acts as the absorber plate. The covered plate collector in its many forms utilises a translucent cover above the absorber plate. Compared with the bare plate collector higher collection efficiencies are obtainable with covered plate collectors but at the expense of increased complexity and cost.

The optimum design suitable for use at farms and mills in developing countries is probably the bare plate collector which is capable of operating at a collection efficiency of 40-50% with an airflow of 0.10 kg/s.m^2. With typical insolation levels in many tropical countries of the order of 0.5 kW/m^2 such collectors are capable of providing mean daily elevations in air temperature of 5-10°C with heat outputs of 0.20-0.25 kW/m^2 of collector area. Covered plate collectors, operating at efficiencies of 60-70%, are capable of providing air temperature elevations of 10-30°C but at a lower airflow.

A major advantage of the bare plate collector is that it can be easily incorporated into the roof of a dryer or storage building. Corrugated iron is a popular and inexpensive roofing material in many areas and when painted black forms an excellent solar absorber. A false ceiling can be fixed to the roof joists so forming a shallow duct running the length of the building and easily connected to a fan via ducting at one end of the building. The heat available from the collector is weather dependent and consideration should therefore be given as to whether solar energy should be the sole source for heating the air or a supplement to more conventional heating systems.

Mechanical Dryers

Batch-in-Bin Dryers

The small capacity version of the batch-in-bin dryer, otherwise known as the flat-bed dryer, has been developed for farm- or village-level use. Its capacity is of the order of 1-3 tonnes/day with drying times of 6-12 hours.

The flat-bed dryer is simple to construct using easily available and inexpensive materials and easy to operate with unskilled labour. The walls of the drying bin can be constructed of wood, brick or metal. The floor of the drying chamber is preferably made from fine wire mesh, suitably supported, or perforated metal. If these are not available then sacking spread over a coarser but stronger wire mesh can be used. To facilitate an even airflow through the bed the length of the drying chamber should be 2-3 times the width. The height of the plenum chamber is of the order of 0.3 m. Unloading ports can be fitted at intervals in the walls of the drying chamber.

In order to prevent excessive moisture gradients through the bed, the depth of grain in the bin is relatively shallow, 0.4-0.7 m. and the air velocity is usually of the order of 0.08-0.15 m/s for maize and 0.15-0.25 m/s for paddy. The temperature of the air is selected according to the desired safe storage moisture content of the grain. For the drying of paddy in tropical areas an air temperature of 40-45°C is usually used, with a heater capable of raising the air temperature by 10-15°C. With such bed depths and air velocities the pressure drop over the bed is relatively low, 250-500 Pa, and therefore simple and inexpensive axial-flow fans can be used. Typically power requirements are 1.5-2.5 kW per tonne of grain for a belt-driven fan powered by a petrol or diesel engine.

Operation of flat-bed dryers invariably results in a moisture gradient between the lower layers and the higher layers of the bed. This problem can be reduced by careful selection of drying temperature and airflow conditions but, even so, gradients of 34% moisture are to be expected. Turning of the grain in flat-bed dryers at intervals can alleviate the problem but this extends the drying time and requires additional labour.

The flat-bed dryer is easily loaded from sacks by hand. However, unloading the dried grain into sacks can be time- and labour-consuming; placement of the drying bin on a tilting frame has been investigated but this incurs additional costs.

Dryers of this type have been developed in many countries and designs are available from the University of the Philippines at Los Baños (UPLB), Los Baños, Philippines and the International Rice Research Institute (IRRI), Manila, Philippines.

IRRI have also developed a vertical batch-in-bin dryer which operates more efficiently than the flat-bed dryer. It differs from the latter in that the airflows horizontally through the bed on either side of the plenum chamber and exhausts through slatted sides. The bin is easily unloaded by removing the slats. Details are available from IRRI.

Both direct and indirect heaters can be used with the flat-bed dryer. Solar air heating can also be an option. The waste heat from the engine used to power the fan can be used. Heating of the air by 5-10°C using waste engine heat is possible but the engine exhaust gases should not be drawn through the grain; the exhaust gases should be ducted outside the housing around the engine. A development of this principle is the moisture extraction unit (MEW) in which the fan is directly driven by the engine and the air is drawn over and around the engine block and exhaust pipe.

Large capacity batch-in-bin (or in-store) dryers can be used in cooler dryer areas. The advantage of this technique is that the bin is used for both drying and storage with savings in both capital and operating costs. With heated air at a temperature of 40-45 °C, bed depths of 2-3 m can be used with air velocities through the bed not exceeding 0.08 m/s. Since drying times to achieve reductions in moisture content of 5-10% can be of the order of 20-40 days, this method should not be used in humid areas with grain of moisture contents greater than 18% because of the risks of sprouting and mould growth in the upper layers of the bed.

Large batch-in-bin dryers are usually round or rectangular and range in capacity from ten to several hundred tonnes. With large bins, air distribution ducts at the base of the bin are used rather than a plenum chamber. To ensure good air distribution through the bed, ducts should be spaced from each other at a distance of half the depth of grain and one quarter the depth from the end and side walls. Air velocity through the ducts should not exceed 5 m/s because of pressure drop factors. More than one fan can be used to provide the airflow required. Detailed information on duct design and airflow distribution is presented by Brooker et al.

In-bin layered drying with ambient air can be performed with confidence in locations where the relative humidity of the air is less than about 70%. An initial layer of grain, 0.6-0.9 m deep, is loaded into a storage bin, 5-10 m deep, and further layers are added as drying proceeds. Over-drying of the grain is minimized because of the low air temperature. In the USA an airflow of 0.025-0.06 m^3/s per tonne was used to dry paddy from

20% moisture to 16% moisture within 14 days when ambient temperature ranged from 18-24°C. Careful and skilled management is required to ensure that each layer is dried before the succeeding layer is loaded into the bin.

Some success has been reported in Indonesia for the drying of paddy from 18% moisture to 13% moisture. The latter also demonstrated that field-wet paddy (24% moisture) could be dried safely in bulk to 18 % moisture by continuous aeration (24 in/day) with ambient air, regardless of the humidity of the air. Subsequent drying to 14% moisture or less was accomplished by drying at times of lesser humidity and with the addition of waste engine heat.

There can be a need to dry grain in sacks in certain instances: for example, at central drying facilities where farmers wish to retain access to their own grain. Stacks of sacks are laid over air distribution ducts with no need for a conventional drying bin. Drying proceeds in much the same manner as for bin drying. After use, the air distribution ducts can be dismantled easily to allow use of the building for other purposes.

Re-circulating Batch Dryers

This type of dryer avoids the problems of moisture gradients experienced with bin dryers by re-circulating the grain during drying. The dryer is a self-contained unit with an annular drying chamber, 500 mm thick, around a central plenum chamber, a fan and heater, and a central auger for transporting the grain from the bottom to the top. When drying is complete the grain is discharged from the top. Most dryers of this type are portable and can be moved relatively easily from farm to farm.

Air temperatures of 60-80°C are employed with air flowrates of 0.9-1.6 m^3/s per tonne of grain, twice that used in flat-bed dryers. However, since the grain is only exposed to the flow of hot air for relatively short times within each cycle, too rapid drying rates are avoided and moisture distribution within individual grains is equalised during the period the grain remains in the non-drying sections at the top and bottom of the dryer. Control of the drying rate can be effected by adjusting the auger speed to regulate the flow of grain through the dryer.

Another version of a re-circulating batch dryer is rectangular with drying chambers on either side of the heater, fan and plenum chamber. Under each drying chamber are horizontal screw conveyors that collect the grain and return it to a screw auger at one end that lifts the grain to a holding section at the top. A screw conveyor in the holding section distributes the grain evenly along each drying chamber.

The capital cost of re-circulating batch dryers is considerably greater than batch-in-bin dryers because of their greater complexity and incorporation of handling and conveying devices. However, throughput is greater due to the shorter drying times and the quality of the dried grain is likely to be higher. Re-circulating batch dryers require specialist skills for construction and trained operators for successful operation and therefore are not generally suitable for operation by small-scale farmers or enterprises.

Continuous-flow Dryers

Continuous-flow dryers can be considered as an extension of re-circulating batch dryers. However, rather than the grain re-circulating from bottom to top, as in the latter, the grain is removed from the bottom, in some systems, cooled, and then conveyed to tempering or storage bins. In their simplest form continuous-flow dryers have a garner (or holding) bin on top of a tall drying compartment. With some dryers a cooling section is employed below the drying compartment in which ambient air is blown through the grain. At the bottom of the dryer is the flow control section that regulates both the circulation of grain through the dryer and its discharge.

There are three categories of continuous-flow dryers based on the way in which grain is exposed to the drying air:

- crossflow, in which the grain moves downward in a column between two perforated metal sheets while the air is forced through the grain horizontally. Dryers of this type are relatively simple and inexpensive, but, unless mixing systems are incorporated, moisture gradients are set up across the bed;
- counterflow, which employs a round bin with an unloading system at the base and an upward air flow. These dryers are relatively efficient since the air exhausts through the wettest grain. Bed depths of up to 3-4 m can be used;
- concurrent flow, which is the reverse of counterflow drying in that the air moves down through the bed. High air temperatures can be used since the air first comes into contact with wet, and sometimes cold, grain. Drying is rapid in the upper layers but slower at the bottom with some tempering action. Bed depths of at least a metre are used;

Probably the most commonly used continuous-flow dryer is the crossflow columnar dryer, which can be classified as non-mixing and mixing types.

In one version of a non-mixing dryer, drying takes place between two parallel screens, 150-250 mm apart on either side of the plenum chamber. The air escapes from the dryer through louvres on either side of the dryer. The flow rate of grain through the dryer is controlled by a regulator gate at the base of the drying column. Since the grain flows plug-like through the drying section the layer of grain closer to the plenum chamber is dried by hotter and drier air than is the grain on the outside. However, mixing is effected to a fair degree when the grain is discharged and conveyed to tempering and storage bins. Air temperatures of 45-55°C and airflows of 2-4 m3/s per tonne of grain are used. Flow problems can be encountered with very wet and dirty grain as the grain may clog. Teter notes that if very wet paddy is to be dried then the grain should be cleaned and also pre-dried to at least 22% moisture before a non-mixing dryer can be used.

In one design of the mixing type of continuous-flow dryer, a baffle system facilitates the mixing of grain and avoids the development of moisture gradients across the drying bed. Higher air temperatures, 60-70°C, can therefore be used without damaging the grain. Unless screens are fitted on the outside of the drying section lower airflows, 1-1.5 m3/s per tonne of grain, have to be used to avoid grain being blown out of the dryer.

Another design of this type is the LSU (Louisiana State University) dryer. In this version the drying section consists of a vertical compartment across which rows of air channels are installed. One end of each channel is open and the other closed. Alternative rows are open to the plenum chamber and intervening rows to the exhaust section. Alternate rows are also offset such that the channel tops divide the moving stream of grain as it descends providing considerable mixing.

Compared with batch-in-bin dryers and re-circulating batch dryers, continuous-flow dryers offer the largest drying capacity. When large volumes of wet grain are to be dried in a single site these are the types to be considered first. They are most commonly used in a multi-pass drying operation. Investment costs are high but because of the large throughputs operating costs per tonne can be lower than the larger batch-in-bin dryers and re-circulating dryers.

In a multi-pass drying system, continuous-flow dryers are used in association with tempering bins. During each pass through the dryer the grain is dried for 15-30 minutes with a reduction in moisture content of 1-3%. Drying at this rate sets up moisture gradients within the individual grains. After each pass the grain is held in a tempering bin where the moisture within the kernel equalises as moisture diffuses from the interior of each kernel to the surface. The combination of rapid drying and tempering is repeated until the desired moisture content is attained. Using this procedure the actual residence time of the grain within the continuous-flow dryer is of the order of 2-3 hours to effect a 10% reduction in moisture. Selection of the number of passes is a compromise between the dryer efficiency, ie fewer passes, and grain quality, ie longer drying time. Tempering periods are usually 424 hours in duration. The tempering bins may be aerated with ambient air to cool the grain with some slight moisture removal.

It is vital that the operation of drying with tempering is carefully planned and managed to ensure maximum throughput and efficiency. This usually means that the plant is operated 24 hours a day with two or more batches of grain being dried at a time. Well trained management and staff are essential.

Drying Operations

Dryeration

Originally developed for use with maize, dryeration is a combination of heated air drying and aeration cooling. In this process a tempering period is employed between a high temperature drying phase and a cooling phase. Whereas less than 1% moisture is removed if cooling is carried out immediately after drying, as much as 2% moisture can be removed if the grain is cooled slowly after tempering. Damage to the grain is reduced and drying efficiency is improved through better utilization of the residual heat in the grain for moisture removal during cooling. Higher air temperatures can be used in the drying phase since the grain is not dried to such a low moisture content.

Two-Stage Drying

Two-stage or combination drying can be used to relieve pressure on drying facilities during peak periods. For example, paddy at moisture contents of less than 18% can be stored for up to 20 days without significant losses either in quantity or quality. In two-stage drying, grain is dried to an intermediate moisture content, 20% moisture for maize, 18% moisture for paddy, as soon as possible using any of the methods described above and then dried instore to the desired final moisture content over several days or weeks with intermittent use of ambient air or air heated by 3-5°C. Research with paddy in the Philippines (Tumambing & Bulaong 1986; Adamczak et al. 1986) has shown that, in addition to increasing throughput of the first stage dryers, there were substantial overall energy savings and no loss of quality compared to drying to 14% moisture in the conventional manner.

Pre-drying Aeration

Work in the Philippines has shown that wet paddy can be maintained in reasonable condition for 3-7 days when aerated with ambient air. By aerating stacks of sacked paddy at a rate of 0.5 m3/s per tonne for eight hours a day, quality could be maintained for nine days during the dry season and two days during the wet season. Aerating in bulk with similar airflows maintained quality for 14 days and three days respectively. The length of time that paddy can remain in aerated storage without deterioration is dependent on the moisture content of the grain and ambient air conditions.

Drying of Parboiled Paddy

After parboiling, paddy contains about 35% moisture. During the parboiling process the starch is gelatinized which confers quite different drying properties to that of field paddy. It has been shown that in the drying of parboiled paddy, significant damage does not occur until the moisture content falls to 16%, regardless of the drying method or the rate of drying,. Cracking then occurs some time after the grain has cooled. The recommended drying procedure is to dry the parboiled paddy to 16-18% moisture as fast as facilities permit, temper it for four hours if warm or eight hours if cooled, and then dry in a second operation to 14% moisture. Air temperatures of 100-120°C can be used for parboiled paddy in continuous-flow dryers.

Drying of Seed Grain

If grain is destined for use as seed then it must be dried in a manner that preserves the viability of the seed. Seed embryos are killed by temperatures greater than 40-42°C and therefore low temperature drying regimes must be used. Seed grain may be dried in any type of dryer provided that it is operated at a low temperature and preferably with greater air flowrates than generally used. It is essential that batches of grain of different varieties are not mixed in any way and therefore the dryers and associated equipment used must be designed for easy cleaning. In this respect simple flat-bed dryers are more suitable than continuous-flow dryers.

Teter noted that seed paddy can be sun dried at depths of up to 30 mm but that the final stages of drying to 12% moisture should be conducted in the shade to avoid overheating and kernel cracking. Flat-bed dryers can be used with bed depths of up to 0.3 m, air temperatures not exceeding 40°C, and airflows of 1.3-1.7 m^3/s per tonne of grain.

Cross-mixing between batches of different varieties can be avoided by drying in sacks in a flat-bed dryer although care must be taken in packing the loaded sacks in the dryer to ensure reasonably even distribution of airflow. Specialised tunnel dryers in which sacks or portable bins are individually placed over openings in the top of the tunnel have been developed.

Novel Dryers and Recent Developments

Fluid Bed Drying

This type of dryer in which individual grains are suspended and sometimes transported by air moving at high velocity, 2-3 m/s, can produce very evenly dried grain. Recent research in the Philippines has indicated that the fluid bed dryer has promising potential for the rapid first-stage drying of paddy to 18% moisture in two-stage drying. Paddy at a bed depth of 100 mm can be dried from 24% to 18% moisture in 15 minutes with air at 100°C and a velocity of 2 m/s, with no adverse effects on quality. However, due to the high air velocities required to fluidise the paddy, power requirements for the fan are high and the thermal efficiency is low compared to conventional (fixed bed) drying. Re-cycling of the exhaust air was identified as a potential means to improve the thermal efficiency.

Conduction Drying

Work at IRRI has investigated the use of a heated floor dryer. This consisted of a metal floor heated to 50-90°C by circulation of water heated by a furnace burning agricultural wastes. Paddy at depths of up to 60 mm could be dried from 22-26% moisture to 18% moisture in 1-2 hours depending on the floor temperature. Frequent raking was necessary but no parboiling effects were recorded and grain breakage was generally lower than that of sun dried paddy. This method is also considered as an option for the first (rapid-drying) stage of two-stage drying.

The Warehouse Dryer

This dryer has been developed for use with a wide range of crops including maize and paddy. Its particular feature is the use of a wind-powered vortex flow inducer as an alternative to conventionally powered fans for generating increased airflow over and around the drying grain. The flow inducer is mounted centrally on the roof of the dryer building and draws air, heated by a furnace and heat exchanger, through the drying bins or trays positioned in the middle of the dryer. The performance is governed by the velocity of the prevailing wind.

Rotary Drying

This method of drying has been researched at the IRRI in Philippines and also at the AIT in Thailand. Small dryers for farm use were developed at the IRRI as reported by Espanto et al. A directly-heated version consists of a perforated iron drum (0.6 m in diameter and 0.9 m long) mounted over a portable stove. The interior of the drum is fitted with flights to facilitate mixing and uniform heat transfer. The drum is rotated manually. An indirect-heated dryer is constructed from a 2001 oil drum also mounted over a stove. Air is passed through the drum by a fan.

The performance of both dryers was very similar. Batches of 25 kg of paddy at 28% moisture can be dried to 18% moisture in 50-60 minutes and batches of maize from 33% moisture to 18% moisture in 80 minutes. Larger versions of the indirectly- heated dryer have been built. With paddy, milling quality was improved relative to sun drying but the viability of the grain was greatly reduced due to the high temperatures attained at the drum surface. Similar results were obtained with the AIT dryer as reported by Jindal & Obaldo.

Microwave and Infrared Dryers

When grain is irradiated by electromagnetic energy high temperature potentials are generated between the interior and surface of individual grains. Moisture therefore migrates to the surface where it evaporates to the surrounding air. The rate of airflow necessary is that required to absorb the moisture and not as the provider of latent heat. This reduction in airflow would minimize the dust and other pollutants discharged to atmosphere. More uniform drying is possible compared with conventional heated-air drying. However the capital cost and energy consumption of the microwave equipment necessary is considerable. Radajewski et al. in Australia investigated, using simulation techniques, the use of microwave heating as a means of pre-heating wheat before drying and concluded that the reduction in drying time could not offset the power consumption required for microwave heating. Infrared heating systems are similarly expensive, and since infrared radiation only penetrates superficially it is necessary to agitate the grain thereby exposing all the surface area to the radiation, thereby incurring additional cost.

Storage for Different Periods

The duration of storage is of vital importance in deciding the most appropriate storage practice. Thus, storage can be classified into the following categories.

Transit Storage

This is the shortest term storage where the grain is being transported from one place to another or where some kind of rotation is practised so that the old stock moves out as the fresh stock comes in. Many of the Government go downs, godowns at seaports and go downs of retailers are examples of transit storage.

Transit storage has been maintained in a variety of sheds available on an emergency basis, from time immemorial. In these sheds, bag storage has generally been practised with the bags arranged in stacks. However, the best method will be to have proper go downs with permanent cubicles in which the bagged grain can be suitably stacked. These cubicles should be constructed so that they can be made air-tight both for storage and fumigation, if need be. This would provide ideal storage conditions and also the necessary handling facilities for bags needed for transit storage.

Short-term Storage

This type of storage is practised by cultivators who generally like to store their seed grain from harvest to sowing and foodgrains from harvest to harvest. Storage structures generally use for short-term are bukhari, kothar, morai, etc., which are examples of non-airtight bulk storage.

From ancient times grain has been stored in India in bulk in mud bins. These earthem structures of various shapes and sizes provide the easiest and most economic methods for the storage of grains under rural conditions in India. However, in this type of storage, one often finds the grains infested with insects. In some places rats also pose a serious problem since they easily cut through the mud walls. In some wet regions of the country, grains are also found to be affected by the high humidity conditions. As a result, the loss of food grains in storage is often quite considerable. With a view to reduce these losses an ideal storage structure has been devised by the Indian Agricultural Research Institute, Pusa Road, New Delhi. It is called the Pusa Bin.

In the Pusa Bin, a thin sheet of polythene film (0.17 - 0.18 mm thick) is embedded in the mud wall of an ordinary earthem structure. The idea behind the sandwiching of the polythene film within the body of the wall is to combine the mechanical strength of the mud wall with the effective imperviousness of polythene films to vapours and gases. Also, this film is impervious enough to oxygen with the result that the oxygen tension within the structures reduced to such an extent that insect multiplication becomes impossible. At the same time, the earthem layers both inside and outside the polythene film keep the film safe from mechanical injuries due to abrasion and handling stress and strain.

Thus, the Pusa Bin combines all three major requirements of safe storage, namely,

(a) it is moisture proof,

(b) it is sufficiently air-tight, and

(c) its walls have poor thermal conductivity.

Long-term Storage

This type of storage is for long periods as required by latge-scale trade stockist and Government agencies desiring to keep buffer stocks or maintain food banks. For long-term storage, careful planning and implementation of the storage practices is required. Prior to the storage of grain, it should be thoroughly dried as the moisture content of the grain is the most vital factor for safe storage. Grain can be dried in the sun but it is highly advisable that a suitable grain dryer be provided for each storage godown.

Air-tight bulk storage is best if the stock has to be maintained on a long-term basis. It is generally in the form of modem silos above ground or as airtight moisture-proof under ground pits. Large-sized structures constructed on the basis of the Pus a bin will also serve well for long-term storage. Such storage structures housed in storage godowns are permanent means of long-term storage. The cost of such storage is quite cheap in the long run. The go downs should have adequate provision for making the whole structure sufficiently air-tight for fumigation and proper aeration after fumigation. This can be easily managed by having ventilation fitted with proper exhaust fans and also a suitable arrangement for closing the ventilator air-tight.

Requirements for Safe Storage

Based on the above mentioned storage considerations we can make the grains less susceptible to insect attack by the use of efficient cleaning and drying equipment and by the application of fumigants and good storage practices. After harvesting and drying, grain and seed should be stored in clean, insect-free, weather-proof storage places from which nearby sources of insect infestation have been eliminated. The sources of insect infestations in stored grains vary with crop and region.

In most cases infestation begins in the field because enormous stocks of grains are stored on the farm either as feed or as market grain. Besides field infestation, infestation in stored grain originates in storage facilities or from nearby stores or accumulation of feed, or other infested dry food products. Therefore, the safe storage of food grains depends upon the storage structure, duration of storage, the local climatic conditions and the quantity to be stored. The storage structures should be of modem construction, easily cleaned and tight enough for fumigation. Every effort should be made to eliminate dead spaces in walls and floors where accumulations of grain offer food and housing for insects. Steel bins that are easy to clean and can be made air-tight are best for storage of small grains. The advantages and disadvantages of air-tight storage are given here.

Advantages

Sealed air-tight storage provides a cheap method of insect pest control. Due to the respiration of the stored product and of any insects present, oxygen is used and carbon-di-oxide is formed, which results in the death of insect pests. In order to accelerate this process, a lit candle can be placed in a tin at the top of the silo just before closing it. The burning candle quickly uses a great deal of the oxygen present. It is important to fill the silo to the top, as the oxygen present is then used up faster. Another main advantage of air-tight storage is the fact that moist outside air cannot enter the silo.

Disadvantages

Important dis-advantage of air-tight storage is that further drying of the produce in the store is impossible. Therefore, the crop needs to be well dried before placing in the store.

Additionally, it is not convenient if the user regularly needs to open the stock to remove food. However, when this happens the whole principle of air-tight storage is lost, because every time the store is opened fresh air enters. It is also extremely difficult to provide regular control of the product through inspection without letting the air inside.

Damage by Insect Pests

Much of the damage to stored grains by insect pests is done directly to the kernels. Their larvae destroy many times their own weight of food during their growth period. Besides, many species feed on the germ of the grain, so that its viability is reduced and germination is impaired or destroyed. They frequently cause grain to heat resulting in a musty odour. Deterioration and rotting of the surface grain sets in as a result of grain heating and consequent moisture migration in storage. Degree of damage depends on three factors namely.

(i) moisture content of the stored grains;

(ii) temperature inside the storage place, and

(iii) oxygen level, besides food supply and human activities.

It has been established that most insects do not thrive below 9 percent moisture content of the grains. So, it is essential that before grains are put into storage, the moisture level should be 8 percent and absorption of the moisture from the air should be prevented. Higher temperature favours the multiplication of insects, hence the inside storage temperature should be maintained low. Oxygen content inside the container under air-tight conditions also has a deciding influence on the infestation of grains and multiplication of insects.

Types of Storage

When storage is for a long period or/and in large quantity, it is referred to a bulk storage and is generally in the form of warehouses, silos above ground or pits underground. Storage of food grains for short periods or/and in small quantity is usually in bags and is known as bag storage.

Bulk Storage

Bulk storage is preferable for wheat, paddy, barley, gram and other coarse grains. To take full advantage of bulk storage the grain should be free from insect infestation, well dried and should be stored in damp-proof and uninfested bins. If infested, grain in bulk can be fumigated at a low cost to prevent further deteriorations and cross-infestation of uninfested stocks. Bulk storage leads to considerable economy in storage space, its maintenance and the cost of gunny bags. After bulking the grain, empty disinfected gunny bags can be re-used.

The bulk grain should be regularly inspected at monthly intervals. Samples should be drawn with the help of thermo-sampler from different levels to know the moisture content, heating, insect infestation, fungus attack and cake formation.

In case of infestation, fumingate the grain with aluminium phosphide at the rate of 3-4 tablets (3 gm each) per tonne, EDCT at the rate of 35 kg/ $100m^3$ and EDB at the rate of 64 gm/tonne with 7 days exposure period. If the commodity shows signs of heating it should be aerated by the aeration arrangement provided in the structure. Insecticides should not be stored near the bulk grain.

Advantages

The striking advantages of bulk storage systems are mentioned here.

i. Minimization of loss due to spillage;

ii. Disinfestation measures are more manageable thus insect losses are minimal and no bird losses;

iii. A centralized system is possible for inspection of food, grains and also for carrying out operations like fumigations, aeration, etc.

iv. The quality can be maintained more easily;

v. Human errors can be avoided to a greater extent;

vi. Processing operations can be planned systematically;

vii. Lesser land costs and

viii. Better utilization of the space.

Disadvaotages

Basically, the following are the constraints in adoption of bulk storage system.

i. Huge initial costs;

ii. Needs modernisation of entire transport system in the country and

iii. Multiplicity of varieties of grain in developing countries.

Bulk storage at farm level

Normally, the design of bulk storage structures depends upon the local climatic conditions, material availability and commodities grown. India, being a vast country with various resources and climatic conditions, a single design cannot cater the requirements of farmers of different regions. Keeping this practical consideration in view, the Indian Grain Storage Institute which is a National Institute dealing exclusively with the storage problems, has designed and tested number of structures. They are classified into two categories.

The first category is indoor and outdoor structures and the second category is metallic and non-metallic. Designs were evolved by combination of these two categories. Some of the designs of each category are given here.

a. *Indoor metallic structures*

- i. Domestic bins;
- ii. Urban bins;
- iii. Spiral lock seam tube bins.

b. *Indoor non-metallic structures*

- i. Pucca kothi;
- ii. Cavity wall bin;
- iii. Plywood bins;

c. *Outdoor metallic structures*

- i. Flat and hopper bottom module bins;
- ii. Corrugated galvanized sheet bin;
- iii. Outdoor flat bottom bin;
- iv. Composite bin.

d. *Outdoor non-metallic structures*

- i. Reinforced cement concrete ring bin;
- ii. Reinforced brick bin;
- iii. Hollow block bin.

However, 29 different designs of metallic storage structures have so far been recommended by the Dept. of Food for adoption.

Bulk storage of high moisture grain

Keeping the special requirements of storage of high moisture grain in view, the Indian Grain Storage Institute had evolved welded wire mesh bin for indoor use and high moisture grain storage bin for outdoor use. In these designs the concept of natural ventilation system (without supplemental heat) was tested successfully. This principle can be further developed and adopted for large scale storage of high moisture grain as storage of high moisture grain is assuming greater importance in developing countries.

Bulk storage at market level

Out of the total production of grain in India, two-third is being stored by the farmers and 25 to 30 million tonnes is handled and stored by the Public agencies and the remaining by traders either in their own stores or in market yards. But bulk storage is not prevalent at traders level and a beginning in this direction can be made from traders level storage.

Bulk storage at cooperative level

The bulk storage at cooperative level has high potential. But there is a problem when cooperative handles the stocks of large number of farmers. The varieties are more and the quantity stored by each farmer is small, in addition to this, each farmer thinks that the grain he produces is superior in quality than others so he has reservations of mixing his grain with others. Introduction of bulk storage at co-operative level is not taken up in our country due to these problems.

Bag Storage

Bag storage is largely practised in the trade go downs, mainly because of the ease in handling and transport. Bagged grain should be stacked on racks, at least 30 cm from the walls of the warehouse and far enough apart to allow inspection and cleaning. Infested bags, if any, can be easily segregated and treated. Bags made of paper, paper laminated to cloth or back- filled fabrics, and cartons of fibre board offer more resistance to insect penetration than ordinary cotton or jute bags.

Although storing grains in bags is not very ideal, it is one of the common mode of storage. Often, the bags themselves serve as a source of infestation. Hence, it is advisable to use new bags or put old sacks in boiling water and dry them in bright sun light before use. Bags of uniform size and weight should be chosen. If there are any holes they should be mended. Fill insect free, clean, cool and dry grain in the bags. Each kind of grain should be put in a separate bag. The mouth of the filled bags should be stitched. In case of machine stitching, five stitches in 2.5 cm should be used. The god own should be kept thoroughly cleaned and sweeping should be collected from time to time, should be sieved, winnowed, cleaned and sound grain recovered should be blended with the rest of the bagged grains. The remnants of the sweeping after recovery of the sound food grains should be disposed of by burning, dumping or for industrial or manurial purpose. The fertilizers or the pesticides should not be stored along with the food grains.

Underground Storage

In the underground storage the grain is more free from the seasonal changes of the temperature and humidity provided seepage does not occur, and is safer from various external sources of damage inducing theft. On the other hand, above ground storage is more convenient for inspection and handling of the grain and can be maintained in more hygienic conditions.

The structure in which food-grain is stored in bags should be constructed on a raised well-drained site not liable to flooding or inundation. It should at least be O.5km away from the places which are a source of infection, like kilns, flour and bone crushing mills, garbage dumping-ground, slaughter houses and tanneries. There should not be any tree near the structure, otherwise birds will be a nuisance. As far as possible, it should be situated near a transport head or main road. The floor of the go downs should be constructed of either cement concrete or store slabs with a slope of 4 cm from the wall to its outer edge to prevent rain water from getting inside the go down through the doors. It should be free from sub-soil moisture. The roof should be water proof. The rain water pipes should be provided for drainage of rain-water from the roof. The structure should be rat-proof.

All the cracks, crevices, holes in walls, floor and ceiling should be filled with cementing material. The structure should be disinfected with the recommended insecticides, namely, Malathion 50% at the rate of 3 litres/ 100 sq. metres, Pyrethrum with 2.5% or Pyrethrin at the dilution of 1:100 at the rate of 3 litres/100 sq. metres. The bags should be stacked 60-70 cm away from the walls with proper alleys around to provide free ventilation of air facilitate periodical inspection of stacks and spraying and fumigation. Each stack should be 9 m x 6 m x 5 m compact and flat on the top.

The maximum permissible height of stacks for major food grains like wheat, barley, oat, maize, gram, pulses, jowar and other millets is 20 layers while in case of rice and milled pulses it is 16 layers.

Storage on Ground

This is for temporary storage, following on immediately from harvesting and lasting only a few days, either because the farmer has not had time to bring in what he has harvested or because he wants to let it dry in this manner for a while if there is no prospect of rain. Unfortunately, during this period, there may be appreciable losses for instance, paddy may be devoured or carried off to burrows by field rats, and ground- nuts, maybe eaten into by bugs of genus Aphanus, which suck the oil from the kernel.

Ventilated Granaries

In this type of granary, the peasant farmers usually store their unthreshed cereals, because in this way, they are afforded better protection against insect and are less susceptible to damage by water infiltration. Rice is generally not threshed but kept in the form of paddy, while groundnuts are kept in their shells. These granaries are cylindrical, or sometimes rectangular. The grain is removed via a small opening in the side, closed by means of a door made of boards or flat metal. The dimensions vary from village to village and their capacity ranges between a tonne and a tonne and a half. These granaries are used for unthreshed cereals, provide good ventilation of the grain and there is little insect infestation of their contents. They are vulnerable to rats, but, to deter them, one should refrain from constructing them under trees.

Silos

Silos are usually constructed of steel or reinforced concrete and comprise high cells of various cross-sections placed side-by-side. They have inlets and hoppers for loading and unloading respectively. Mechanical management is also provided for loading and unloading in case of large capacity silos.

The basic layout of vertical silo installation will have the following components.

i. *Reception pit:* The material brought to the complex are tipped into this.

ii. *Elevator:* For raising the material from pit or the top conveyor of silo for loading into silos.

iii. *Loading belt conveyor*: For loading into the silos a belt conveyor is used which ins installed on the top.

iv. *Unloading belt conveyor:* For unloading from silos, the belt conveyor is used which is installed below the outlet of the hopper bottom.

v. *Hopper bottom*: Large scale silos are provided with hopper bottom. The hopper bottom should ensure easy gravity flow.

Principal types of silos

Broadly, there are the following three types of silos.

i. Reinforced concrete silos: They can be extremely large having height of 50, 75 and even 100 metres.

ii. Cylindrical silos with corrugated steel sheets:

These silos are made of either corrugated or flat galvanized steel sheets and corrugations are kept horizontal while in flat ones, the thickness of the sheet depends on the capacity of silos. They are pre-fabricated for easy erection at site.

In the developing countries like India, the silos commonly in use are briefly mentioned here.

Steel Silos

In India, bulk storage metallic silos were erected in the year 1958, in Hapur, Uttar Pradesh. Each of these silos are about 6.7 m in diameter and 23 m in height having a capacity of 500 tonnes. There is a head house having a diameter of 8.5 m and height of 47.3 m. Head houses contain grain elevators, auxillary tanks, grain cleaning and disinfecting equipment, grain drier and weighing machines. A drying arrangement for reduction of moisture content is also provided. There are provisions for recording temperature fluctuations in the silos aeration and fumigation.

Flat storage plant

This plant is having a capacity of 55,000 million tonnes. There are two elevators, six screw type conveyors. They have provisions of temperature recording and aeration.

Concrete bins

Concrete bins of capacities 2000 and 4000 tonnes each constructed in India in 1965. Subsequently, more bins are added to increase the capacity of this type of storage. These are constructed in Madhya Pradesh and Maharashtra.

Horizontal silos

The horizontal silos have large diameter and lesser height. These silos are preferred if the ground area available is large and the height of the handling equipment is to be smaller. But normally vertical silos are preferred than the horizontal silos as they are economical.

Hexagonal silos

The capacities of this kind of structures may vary between 500 tonnes to 2000 tonnes. Here, the walls are constructed with bricks and cement mortar having thickness of 400 mm. The hopper bottom is provided with reinforced cement concrete which is 150 mm thick. The whole structure is supported by concrete columns.

Factors Affecting Grains in Storage

This has been observed that quantitative and qualitative -losses in stored food grains may occur due to physical (abiotic), biological (biotic), chemical (breakdown of produce and pesticides) and engineering (structural and mechanical, aspects) factors. Grain temperature and moisture are the two important factors which affect the rate of metabolism, growth, development, reproduction and general behaviour of stored grain insect pests and fungi. The main factors affecting the storability of food grains are mentioned here.

Temperature

This is the most important ecological condition required for grain storage.

Effect on insect activity

Temperature is one of the most important extrinsic factor which exerts a profound impact on the rate of metabolism, growth, development, reproduction, general behaviour and distribution of pests.

The lower temperature at which the insects are able to develop lies between 15.5 and 18.30 centigrade. The optimum temperature for most of the insects lies between 29.4 and 32.20 centigrade. The optimum temperature for *T. granarium* and *R. dominica* is probably the maximum amongst the storage insect pests.

Effect on micro-organisms

The micro-organisms differ in their thermal requirements for growth and are commonly classified as Psychrophiles, Mesophiles, and Therrnophiles. Their required temperature for growth -minimum, optimum and maximum are: Psychrophiles : 8 to 0°, 10 to 20°, and 25° to 30°C, Mesophiles : 5° to 25°, 25° to 40°, and 40 to 45°C, Therrnophiles : 25° to 40°, 50° to 600, and 70° to 90°C.

Micro-organisms grow faster and sporulate soon as temperature approaches their optimum. Most of the fungi and bacteria die within 10 minutes at 55°C while few servive at 55°C for some length of time. Dry spores of many fungi may survive at 87°C for 30 minutes while only a few survive at 121°C for the same length of time. Bacterial spores which are more resistant to moist heat than fungal spores may withstand 95° for 45 minutes or even 100°C for as long as 20 hours.

Effect on mites

Under Indian conditions some of the mites develop faster if the temperature is 18° to 22° Centigrade.

Effect on respiration

Respiration, whether of seeds, moulds or insects depends upon chemical reactions and is, therefore, accelerated by an increase in temperature until it is limited by such factors as the thermal inactivation of enzymes which are involved, exhaustion of the substrate, limitation of oxygen supply or accumulation of inhibitory concentration of carbon-di-oxide. It has been observed that the rate of respiration of grain increases with the temperature until inhibitions of vital process begin.

Moisture

In food grains there is a fairly well-defined relationship between moisture content of grain and relative humidity of the environment. The food grain moisture is due course tends to come in equilibrium with the humidity of the air. Thus, the amount of moisture fluctuates according to the temperature and relative humidity of the surrounding air. The rate at which the grain will lose or gain the moisture generally depends upon its condition, type of grain and extent to which it is exposed. However, the moisture keeps varying according to the changes in the environment. In India, the grain is exposed to the environment when it is spread in thin layers at the farmers houses resulting in the loss of moisture to a greater extent. This depends upon the environmental conditions at that time and also the thickness of heaps or layers of the grain.

Effect on insect activity

For majority of storage insect pests, moisture content below 10 % is not adequate for normal activity and development. Insect infestation tends to increase with increase of moisture content above 10% up to a certain limit. Generally, insect activity is rapid at relative humidities in excess of 75 percent. The corresponding moisture content of wheat is 14 to 15 percent.

One of the most remarkable feature about the storage insect pests is that some of them can live in products having a very low moisture content. However, a certain minimum moisture level is needed and this varies considerably for different species. The minimum relative humidity of *Trogoderma granarium* everts, *Tribolium confusum*, Tribolium *castaneum*, *Callosobruchus analis*, *C. rhodesianus* has been accorded as low as percent. At this minimum level development is frequently very slow and mortality is high. Other conditions being favourable, the rate of development generally increases as moisture content of food rises. Optimal conditions are largely obtained when the moisture of food reaches about 14 percent. At this moisture level, other deteriorating changes also set in and this, therefore, is considered the highest level of moisture for safe storage of grains.

Insects that live in food having a low moisture content exhibit a remarkable ability to consume and utilize water and are certainly well adapted for the products in which they live. It has been known that weevils are attracted to moisture. The moisture content has to be in excess of 12% if rapid multiplication of weevils is required.

Effect on micro-organisms

Water contributes to the chemical and physical structures of micro-organisms and is involved in cellular work of metabolism and translocation of material. Moisture has a great affect on development of fungi, bacteria and yeast, etc. Micro-organisms are classified as hydrophytes (minimum requirements is 90% or more relative humidity (RH), mesophytes (minimum requirements is 80 to 90% RH) and xero-phytes (less than 80% RH) on the basis of their minimum moisture requirement for growth. The minimum requirements of yeast like fungi is from 88 to 96% relative humidity. The minimum water requirement for spore germination ranges from 62 to 99% relative humidity.

Bacteria requires higher moisture level for growth in grains and grain products.

Hence, the micro-organisms cause discolouration, loss in viability, heating and induct bio-chemical changes besides production of various toxins which are very harmful to human beings.

Aeration of Grains

Aeration is a process of cooling a grain bulk by passing large quantities of untreated air of suitable temperature and humidity, so that the grain in storage remains in an acceptable condition for long periods. Aeration reduces grain temperatures to minimize mould growth and insect/mite activity. It maintains the grain quality and provides more uniform grain temperatures to prevent convection air movement and moisture transfer. Fumigants may also be distributed by aeration fans.

Stored grains has relatively low thermal conductivity. Daily temperature changes have a very small effect upon grain temperatures, and even seasonal changes from summer to winter are likely to cause pronounced grain temperature variation only in most exposed part of a large grain bulk. The aeration equalizes the temperature of the grain thus arresting multiplication of insect and mould activities as well as moisture migration. Grain aeration is more properly considered as a preventive rather than a remedial measure.

Effect of CO_2 and O_2 Concentration

Oxygen and carbon-di-oxide of the inter-granular environment influences the respiration of the grain and consequently also the rate of deterioration and heating. Majority of the fungi are storage aerobes. They fail to sporulate, their spores fail to germinate and their mycelium fails to grow when oxygen concentration is below a minimum which is well adequate for yeast growth.

Effect of Grain

Grain as a component of the environment may influence an organism's chance to survive and multiply by modifying its fecundity, longevity or the speed of development.

Effect of Insect Population Density

Continuous multiplication of insects in stored products makes them increasingly less suitable for human consumption. Under favourable conditions certain insects increase almost to unbelievable numbers.

Light

Generally, the stored product insects seek secluded places and keep well hidden. Stored product insects reaction to light is more varied. Few are attracted towards light while most of the insects avoid it preferring a subdued light or darkness. Different stages of the same species react differently. A stack of gunny bags kept in dark places and most concealed parts, are found to be more' severely attacked than those which are exposed to light.

Susceptibility of Storage Pests to Fumigants

It is already known that the susceptibility of stored grain insect pests to fumigants is influenced by various factors viz., temperature, carbon-dioxide, oxygen, nutrition, population density, pre-fumigation starvation, post-fumigation starvation, relative humidity, sub-lethal fumigation, repeated fumigation, presence or absence and nature of commodities to be fumigated, affect of age, sex of insects, respiration etc. Hence, the success or failure of fumigation operations depends upon these factors.

Respiration of Grains

Production of heat during storage is responsible for enormous waste in stocks is due to respiration of grains, moulds and insects. The combined respiratory activities of these factors are doubtless responsible for primary heating of stored grains. These activities come to an end at about 550 centi- grade. However, some auto-exhaustion may also take place afterwards. Thus, it has been stated that the factors like moisture, temperature, aeration, oxygen, age of the grain, consistency and condition of the grains, micro-organisms, etc. play an important role in the respiration of the grain.

Cracked Grains

Cracked grain and foreign material in excessive amounts can also be considered to be an important factor in storage, specially as for providing favourable conditions for the development of the non-boring type of stored grain insect popularly known as bran and fungus beetles. These do not develop readily in clean grain but feed primarily on grain dust, broken kernels, moulds, etc. It is extremely difficult to fumigate grain which has a high percentage of broken grains and foreign materials.

Sources of Infestations

There are many ways by which stored products may become infested. They include placing clean products in the infested storage, placing infested roducts in clean storage, by mixing infested with uninfested materials, through the use of already infested container, and through the use of infested transportation facilities such as trucks, etc. Infestation may, also arise from natural sources. These may include nests of birds and rodents, the hiding and bedding places of animals, etc.

Integrated Pest Management in the Control of Storage Insects

The characteristic problems of stored-grains pest control in the tropics, with particular reference to developing countries, are concisely described by Taylor, Golob and Hodges. They indicate the crucial importance of effective pest control for a broad range of storage entrepreneurs, including resource-poor farmers, commercial operators, and national marketing organisations; and a wide variety of storage situations in which simple, traditional on-farm systems and large scale bag-storage systems predominate. Bulk storage occurs quite commonly for small bulks, at farm level; but is less common than bag-storage in large scale operations except at large grain mills and at some central marketing depots.

Pest control measures, in general, have to be integrated into an operational system, be it large or small in scale, if they are to be effectively applied. This is a basic principle, not a novel concept, but it connects well with the modern idea of Integrated Pest Management (IPM). The use, in that term, of the word 'management' is appropriate, especially with regard to the need for the integration of pest control measures into management systems, but it should be remembered that the two words, 'management' and 'control', are almost synonymous. The fundamentally important emphasis should be placed upon the word 'integrated'.

Integrated pest control can be defined as the acceptable use of practicable measures to minimise, cost-effectively. the losses caused by pests in a particular management system. For the measures to be cost-effective they must be appropriate to and acceptable into that system. They may be simple or complex but they must suit the system objectives and its technical capabilities. Furthermore, in this context, cost-effectiveness requires that all costs and benefits, including sociological and environmental effects, should have been taken into account.

The term integrated pest management is used to imply that a flexible and technically informed approach is also required. In defining this term it may be considered necessary to specify the inclusion of scientific and cost-effective pest monitoring procedures which permit judicious adjustments to the timing, choice and intensity of control actions. It may also be advisable to point out that specific pest control measures, as distinct from general crop or commodity husbandry practices, should generally be omitted unless the circumstances warrant and permit their cost-effective inclusion.

Insect pest management for stored grains, like preharvest pest management, can thus be seen, historically, as a traditional approach in which good husbandry is the primary requirement. Unfortunately, it must also be acknowledged that the advent of readily available, relatively inexpensive synthetic insecticides has led to considerable over-dependence upon these hazardous tools with, in some cases, a consequent neglect of basic good husbandry. The current emphasis upon integrated pest management is, in effect, a reassertion of the need to put traditional good husbandry in place as the fundamental basis of pest control. In grain storage, as with other durable agricultural products, it is good commodity management and good store management which are the major prerequisites.

The various options for more intensive insect pest control include several which are themselves based upon traditional concepts of pest management. Thermal disinfestation, cooling and hermetic storage are examples. These latter two methods are also examples of the opportunities, provided by the process of storage, to manage the generally enclosed storage environment in such a way that insect pests are prevented from multiplying or, as in efficient hermetic storage, effectively eliminated. Preharvest problems of insect pest control are rarely, if ever, so easily managed!

Control of the storage environment is thus an essential element in grain storage pest management. It involves, primarily, the controls on in-store climate and infestation-pressure which can be achieved by technically sound store design and construction. Equally important, however, is the climatic control attainable by scientific management of the commodity to ensure that the stored grain is itself both dry and cool when loaded or, in ventilated stores and bins with aeration equipment, that the storage procedure achieves drying and cooling sufficiently rapidly. In a fully loaded store it is the stored grain itself which largely determines and stabilises the temperature and humidity conditions in the store.

Commodity management can also control, to a considerable extent, the initial insect infestation level in the stored grain. However, in tropical countries, where preharvest infestation by storage insects is hardly ever completely preventable, the ideal of loading insect-free grain into the store is not often attainable. Special facilities to completely disinfest the grain before loading may not prove cost-effective. The common alternatives, if early disinfestation is required, are to treat the grain, at intake, with a suitable admixed insecticide or to disinfest the loaded grain by in-store fumigation.

Control of grain quality before storage, to minimise the intake of heavily infested and badly damaged or uncleaned grain, is feasible and is commonly practiced to a considerable extent. Even at the small farm level it is possible to segregate the crop at harvest, especially with maize on the cob and unthreshed sorghum and millet, selecting relatively undamaged material with good storage potential and setting aside the more evidently infested or otherwise damaged material which, if there is no other option, can at least be used first. By such means, the rate of deterioration due to insect infestation can be considerably retarded in the main stock of stored grain. There is little doubt that some subsistence farmers use this form of commodity management fairly effectively. Certainly, one can sometimes observe onfarm grain stocks, that have received no special insecticidal treatment, with relatively little insect damage after several months storage at an ambient temperature that would permit the rapid increase of any well-established initial insect population.

Scientific approaches to grain storage pest management, having regard to grain storage as a part of the food production and distribution management system, have sometimes referred to the biological ecosystem concept as a means of comprehending grain storage processes and problems. In a recent contribution Dunkel has applied ecosystem principles in a broad analysis directed towards an improved understanding of physical and biological interactions including socio-economic factors. The purpose was to generate improved understanding of the stored grain ecosystem and to identify objectives for future postharvest research. This treatment of the subject should serve to enhance the growing awareness of storage as a system within a system and to stimulate systematic and objective analysis of grain storage problems. Whether or not one prefers to use the term 'ecosystem' is less important and this somewhat academic term should not be allowed to obscure the main issues.

The integration of various control techniques, within the framework of integrated pest management, has become a focus for research in stored products work. The importance of a multidisciplinary approach to stored grain research has also been stressed. This is very valid but it is useful to recall that a great deal of the research done in the past has been of this nature. It was pointed out, at a stored products pest management symposium held in 1978, that the need for an interdisciplinary approach is generally well-known. Entomology, mycology, chemistry, engineering and food science are commonly involved, but effective integration of technical solutions is often lacking; possibly because some of the more pragmatic disciplines, notably economics, sociology and business management, are not always sufficiently involved. It is the interface between the research team and the storage managers, whether these be individual farmers or a large storage organisation, which may sometimes be the most crucial barrier to progress.

A clear perception of the need for solutions which can be integrated into the management system, because they meet the business objectives and can be accommodated within existing management capabilities, is probably the most important requirement. In this context it is of some interest, although not very surprising, that a correlation has been found amongst farmers in India between 'grain hoarding capacity' (which relates to the farmers' existing business objectives and capabilities) and the adoption of improved storage practices.

Modern theories of pest management have also generated the concept of economic control thresholds (ECTs). An ECT is most simply defined as the level of pest damage which justifies, in cost/benefit terms, the expenditure of resources upon control actions. It is always a variable threshold because the costs and benefits of any action will depend upon the situation and its circumstances. An ECT is situationspecific. This is especially true, and not only for postharvest pest control, when one considers the great differences in opportunities and constraints between the small farm level in developing countries and more sophisticated levels of operation. Nevertheless, it is possible to generalise to some extent For insect control in grain storage the ECT is likely to be at or very close to zero:

a. where consumer demands place a high value on freedom from insect damage and/or freedom from any sign of insect infestation;

b. where there is a definite intention to store grain for a protracted period, in which substantial insect damage can be predicted, and where the eventual market is not insensitive to loss of quality.

In both cases the assumptions are made that at least one potentially cost-effective package of control measures is available and that the package provides for sustained control: including efficient prevention or control of reinfestation. In most sophisticated situations, where consumer quality standards are likely to be high, these will be valid assumptions. Otherwise the ECT could not be zero and the required quality standards would have to be met by reconditioning the product. The losses, including reconditioning costs, would then have to be borne by the system or the consumer. A real economic loss would have occurred, masked by marketing tactics. Such events do, undoubtedly, take place.

Where grain is being stored for domestic use, or for eventual sale in an uncritical market, the ECT may be well above zero. This can be true even when all parameters of economic damage are considered because, in fact, the true economic significance of a low percentage of insect damaged grain may be virtually nil. The actual loss in real food value is likely to be negligible and the market value of the associated weight loss, if saved, may not offset the full costs of pest control actions.

The need to minimise the cost of insect pest control is a major factor militating against the extensive use of some of the more 'environmentally friendly' measures which might otherwise be preferred to the continued use of suitable fumigants and contact insecticides. However, it must also be acknowledged that chemical pest control measures, in grain storage, are often not only the cheapest but also the most reliably efficacious of the possible options. Where this remains true, and for as long as such treatments are accepted as generally safe, one important requirement for IPM in grain storage will be to ensure that chemical measures are recommended only where they can be safely and efficiently used and only when they are economically justifiable.

The economic justification of control measures, including where necessary the use of chemical pesticides, should take account of all costs and benefits. Many of these may be assessable only in subjective terms that are greatly dependent upon local attitudes and sensibilities. However, measurable losses of quantity and certain quality parameters can be objectively determined. A manual of methods for the evaluation of postharvest losses is available and provides useful information, subject to a need for modification of some methods in particular circumstances. A critical review of the methodology gives further guidance based upon experience gained, since 1978, from field work in many developing countries. Choices among methods, several of which have become subjects of considerable controversy, should always be made with due regard to the actual circumstances and the prevailing objective. There is no single 'best method' for all circumstances and, for practical purposes, operational facility and repeatability are generally more important that fine precision. From the storekeeper's viewpoint a demonstrable loss reduction from, say, 5% to 4%, however statistically significant, may be of little or no practical importance; whereas a reduction from 10% to 5% (or from 2% to 1% in a sophisticated storage system) may be of considerable interest: provided, always, that the demonstrated reduction can be achieved in routine practice.

Costs and Incentives to Store

Both producers and consumers benefit from stable prices, which reduce the uncertainties associated with planning farm investment and household expenditure. However storage involves costs, and the only way in which these costs can be recuperated is through a price spread. If storage is to be profitable, people who store grain must receive a price on sale which at least covers the costs of storing the grain since harvest. These include:

- The cost of the store itself (often a rental cost);
- Labour and supervision;
- Pest control;
- Storage and spillage losses; and
- Cost of capital invested in the grain.

In practice, the costs of storage depend on the commodity stored, on the type of storage system, and on unpredictable and variable factors such as pest incidence and climatic conditions. Storage costs also depend on the circumstances of the person, the business or the institution who is storing. The most variable component of storage costs is the cost of capital. For a small farmer or trader, capital may be scarce and costly, and their only access to loans may be from money lenders charging rates of 10% or more per month. On the other hand a Government Marketing Board may have preferential access to loans at low interest, at rates of as low as 10% per annum. There is, therefore, no single cost of storage.

References

Al Hassan, R.M. (1989) *Economic evaluation of storage of grains in the Northern Region of Ghana in Effective post-harvest management of cereals and legumes in Ghana*. University of Ghana, Jan 1989.

Giga D.P. and Katerere Y. (1986) *Rural grain storage in Zimbabwe: Problems, losses and prevention.* ENDA Zimbabwe and University of Zimbabwe, Harare. 96 pp.

Greeley, M. (1987) *Postharvest Losses, Technology and Employment. The Case of Rice in Bangladesh.* Boulder, Coloado: Westview Press, xviii, 345 pp.

Greeley, M. (1991) P*ostharvest Technologies: Implication for Food Policy Analysis.* EDI Development Policy Case Series, Analytical Case Studies No.7. Washington: World Bank

Phillips R. (1981) *Viable Grain Storage for Small Farmers in Developing Countries.* Manhattan, Kansas: Food and Feed Grain Institute, 43 pp.

4

Grading of Foodgrains

Most countries have developed national standards for their main grain crops. These have evolved to facilitate the movement of grain, providing both sellers and purchasers with guidelines to support financial transactions, and ensuring that quality will meet up with enduse requirements.

Where trading involves direct choice and price negotiation in front of the commodity, grading standards are rarely employed; quality is assessed visually and is influenced by enduse, and the price is determined morc by local rather than national factors. For transactions that involve the movement of large volumes of grain over long distances, the buyer may never meet the seller or be able to examine the whole consignment. The standard will provide an unambiguous description of the quality of the consignment and assist in the formation of a legally-binding contract. Standards can also be seen to protect consumers rights through setting limits to the amount of unsuitable or noxious material.

The use of grading standards can send a clear indication of quality requirements to both producer and end-user. Although some countries have sought to support small farmers through purchase of all grain at the same price without regard to quality: under these circumstances grading standards cease to be operative by default. This may stimulate productivity but creates problems for end-users such as millers who require uniformity and consistency in quality to ensure efficient and cost-effective processing.

Whilst establishment of standards can set the guide-lines and rules for sale and purchase of grains, there has to be an institutional framework for their implementation. This is much easier to establish at centres of aggregation of grain e.g. ports, parastatal grain depots, than in the more diffuse rural areas and markets, where control and supervision of regulations is difficult.

Notwithstanding these problems, the establishment of quality and grading standards for producers and users can be beneficial in the following ways:

- Graded grains are likely to be more equably priced than non-standardised grains. This will bring stability not only to market prices but also to the quality offered.
- Prices quoted against a recognised grade assist producers and traders to market their products. This will also benefit net consumers of grain in more stable prices with assured quality.
- Greater conformity in quality through standardisation will provide the millers, bakers and other processors with the consistency necessary for optimum performance.
- Standards reveal clear variations in quality and indicate the opportunities for improvement and the potential rewards to be obtained.
- The sanitary hazards associated with the inter-country movement of grain can be reduced if clearly-defined standards are enforced, particularly in relation to the prevention of spread of serious storage pests like the Larger Grain Borer.

However, the use of standards can have its disadvantages, namely:

- National standards may reflect local end-uses and hinder export to areas that have differing requirements. Whilst restricting opportunities for commercial trade, this can also be detrimental for regional food security. The stimulation of regional trade would require inter-country conformity in export standards. These may have to be more stringent in relation to physical conditions such as moisture content, defective grains and insect infestation, than that for national use.
- The establishment of standards and the quality assurance practices to regulate and enforce them carries costs which have to be carefully considered to avoid imposing unnecessary expense for little improvement in quality.

Quality Characteristics of Grains

Consumers have become accustomed over the years to demanding grain with particular qualities. Where consumers are close to the source of the grain, e.g. in local markets, their own preferences and the laws of supply and demand will control the quality of the grain. However, where grain is traded over large distances, particularly internationally, the consumer will have no direct influence over quality, and regulatory standards must be established and imposed to protect consumer rights. Therefore criteria of grain quality must be established and accepted by all parties in the grain trade. The criteria assigned to grain are the intrinsic varietal qualities and those which are environment- or process induced.

Intrinsic Qualities

(i) *Colour.* Cereal grains are pigmented and range through the colour spectrum from very light tan or almost white, to black. Where extractive milling is required, highly-pigmented varieties may give low yields of white flour.

(ii) *Composition.* Composition, e.g. protein, carbohydrate, lipids and their breakdown products, qualitatively influences product acceptability, by affecting texture and taste. Quality changes evolve slowly in stored grain and more rapidly in milled or processed intermediary products.Some grain components, for example husk, are inedible and quantitatively influence product yield and gross nutrient available to the consumer.

(iii) *Bulk Density.* Each type or variety of grain when in optimum health, fully mature, etc. has a characteristic bulk density. This is defined as the weight per standard volume measured in a standard manner. The same characteristic is variously known as 'test weight', 'bushel weight' or 'specific weight'. If the bulk density varies the trend is usually downwards and indicative of reduced overall quality of the grain. Hence it is often measured in the grain trade. Factors which commonly affect bulk density are insect infestation, excessive foreign matter and high percentage moisture content. In wheat, bulk density is considered to be a reasonable indicator of milling yield.

(iv) *Odour, aroma.* Most grain types, when fresh, have a distinctive natural odour or aroma. This is generally accepted as an indicator of good quality, although some people prefer grain which smells 'old' or even fermented.As with most natural produce, some grain varieties are better-liked than others because of their odour. Certain cultivars of rice, for example, possess aromatic qualities which are considered desirable by some consumers.

(v) *Size, shape.* Rice, as a whole-grain food, is classified by size (length) and shape (length:breadth ratio). Other grains also have size considered in their specification. In general a small range in size assists with processing and handling.

Induced Qualities

(i) *Age.* During the post-harvest phase, grain undergoes complex biochemical changes termed 'aging'. Changes to carbohydrate, lipids and protein fractions result in, for example, firming of texture in rice on cooking, and increased gas-retention capability in wheat flour. For most consumers, the effects of these changes are considered to be desirable. When plotting consumer acceptability of a grain product against its age since harvesting, generally it is considered to be maturing during the upward curve of the graph, and deteriorates only when the curve changes direction downwards.

(ii) *Broken grain.* Grain is marketed normally in whole grain form and is considered to be of inferior quality if broken. Breakage may occur from fissures as a result of excessive drying/weathering conditions in the field or during handling. Breakage reduces quality by reducing acceptability and by increasing susceptibility to infestation during storage. This affects milling yield by contributing to weight loss.

(iii) *Chalky or immature grain.* Empty grains result from sterility and pre-harvest infections and insect attack. Immature grain content is affected by time of harvest. In rice, immature grains are greenish in colour. Thin white (usually opaque) grains are caused by incomplete grain filling and may result from pests or disease. Chalkiness is caused by incompletely filled starchy endosperm which disrupts light transmission, causing opaque regions. In most cereals, chalky areas have lower mechanical strength on crush tests and may break during handling. The broken portion is more easily invaded by certain storage pests.

(iv) *Foreign matter*. Dilution of the prime product by foreign matter reduces the value, and also may affect handling and processing. Foreign matter may be subclassified as:animal origin - insects and their products, rodent excrete, etc; vegetable origin - straw, weeds, seeds, dust, micro-organisms/toxins; mineral origin - stones, mud, dust, glass, metals, oil products, pesticide residues. Elements from all three subclasses may render the grain unfit for consumption. Potentially the greatest threat to health probably is from micro-contamination with the bacterial products of poor sanitation, and with toxins and chemical pesticide residues.

(v) *Infested, infected grain*. Grain mass, and therefore yield, is reduced by infestation. Contamination not only has direct food hygiene implications but also indirect ones, as invading micro-organisms may produce toxins under certain conditions which may lead to acute or chronic illness.

(vi) *Mixed varieties*. A mixture is an indication of poor pre- and post-harvest management and supervision, e.g. seed selection, lot segregation and treatment, contamination, etc. Grains differing in size and other characteristics affect processing potential. Whilst preference for a particular variety may be influential nationally or regionally, internationally-traded grain is recognised usually by grain type rather than by variety e.g. yellow or white maize. Exceptions do occur, e.g. basmati rice.

(vii) *Moisture content*. Moisture content (me) of grain plays a crucial role in post-harvest processing and is associated with most of the induced characteristics. Water vapour will diffuse throughout a bulk of grain and the mc will tend to equalise. 'Hot spots' may occur at a site of increased respiration (caused by sprouting, infestation or microbial activity), and condensation may occur on cold grain or containers.

Grain Standards

The term 'standard' used in the present context refers to the measures that serve as a basis for making comparisons or judging the accuracy of unknown samples. Three types of standard will be covered in this chapter:

- standard *specification* which define and specify a subject,
- standard *test method* by which a specification is tested,

- *grading* standard which allow a subject to be classified into more than one category.

Standards are established for a variety of purposes but mainly: a) for produce grading in agricultural marketing, or, increasingly, b) for the protection of consumers. The requirements of the two groups are not necessarily compatible.

Standard Specification for Grain

There are at least 330 specifications for cereals and cereal products at national and international level (over 50 countries or regions) of which at least 12 are applicable globally. Standard specifications provide criteria to characterize the nature of a commodity, usually on a pass or fail basis.

Most countries have a national standards institution which may issue specifications for commodities as well as methods of testing. Many countries adopt or modify international standards, e.g. International Organization for Standardization (ISO) standards, into their national system. Another source is the Codex Alimentarius Commission (Codex) which operates a committee to formulate standards on cereals, pulses and legumes.

Table 1 shows the ISO standard specification for wheat: a sample is judged against the standard, and may be referred to as wheat if it passes all the criteria listed. The quoted international standards allow 15.5%, whereas the European Community (EC) and Ethiopian standards have lower values, 14.5 and 14.0% respectively.

In the cited examples, not only do we have the potential for mix-interpretation but also, more seriously, the basis for deterioration of the product. For example, take the case of maize with 15.0% mc loaded into a ship in northern latitudes at 4°C. Calculated from data presented by Foster (1982), such grain would have an inter-grain equilibrium relative humidity (ERM) of approximately 65 % which is normally regarded as safe. However, if the grain is unloaded in a tropical country with an ambient temperature of 32°C, its inter-grain ERH becomes 75% - too high for safe storage.

Boxall and Gough monitored shipments of food-aid maize grain from north America to southern Africa. They reported that heating and mould-damage took place when the grain was stacked at port of discharge, and confirmed that the standard mc of 15.5% at loading was too high for the conditions at the destination.

Table 1. International Standard - Wheat - Specification ISO 7970: 1989

Maximum tolerances (% m/m):	
Foreign or deteriorated odour, additives, toxic substances	0
Pesticide residues, other contaminants	National limit or Codex limit
Living insects	0
Moisture content	15.5*
Bulk density, minimum (kg/hl)	70
Damaged grain	15
of which:	
Broken grain	7
Shrivelled grain	8
Unsound grain	1
Grain attacked by pests	2
Other cereals	3
Extraneous matter	2
Inorganic material	0.5
Harmful and/or toxic seeds, bunted grains and ergot	0.5
Ergot	0.05
Falling number, minimum	160

Lower moisture contents are required for certain destinations, in relation to the climate, and duration of transport and of storage.

Source: International Organisation for Standardisation

Table 2. Codex Standard for Maize (Corn) Codex Stan 153-1985

Maximum tolerances (% m/m):	
Abnormal or foreign odour	0
Moisture content	15.5
Blemished grain	7
of which:	
diseased grain	0.5
Broken kernels	6
Other grains	2
Foreign matter	2
of which:	
inorganic matter	0.5
Filth	0.1
Toxic or noxious seed, heavy metals, microorganisms or poisonous or deleterious substances	free from amounts hazardous to health

Source: Codex Alimentarius Commission

A second shipment of grain dried specifically to approximately 14.5% mc suffered damage as well. In comparison, an intraAfrican importation with 12.2% mc did not suffer heating or deterioration on storage. Studies are continuing at NRI on the importance of integrating quality standards between suppliers and consumers.

The problems associated with moisture content may be harmful to international trade in grain and need to be addressed adequately by the standardisation institutions.

Standard Test Methods

There are at least 420 standard test methods for cereals and cereal products at national and international level (over 50 countries or regions) of which at least 75 are applicable globally.

As with the specification of cereals, many countries modify or adopt international standards, e.g. International Organization for Standardization (ISO) standards, into their national system. Another important organisation, particularly for the development of testing methods, is the International Association for Cereal Science and Technology (ICC). Other organisations issuing standard test methods include the Association of Official Analytical Chemists (AOAC) and the American Association of Cereal Chemists (AACC).

Table 3 shows the world production and trade figures for grain for the year 1990. Total grain movement across national borders was approximately 225 million tonnes, representing some 12% of grain production. This large trade in grain highlights the need for both national and international standards to ensure uniformity in quality and quantity of grain.

Table 3. World Grain Statistics

Year: 1990	*Production'000 t*	*Imports'000 t*	*Exports7000 t*
Africa	87751	27785	2725
N + C America	402095	16270	115770
S America	68503	8517	11227
Asia	859650	98878	15808
Europe	283238	40078	62527
Oceania	24587	787	14802
USSR	228854	32862	1641
World	1954678	225177	224500

Source: FAO Trade and Production Year-book

National Requirements

It could be argued that a commodity standard should be country-specific, containing factors such as percentage brokers, foreign matter, moisture content that reflect the types of end-use, be it for commercial or domestic purposes. Government policy on the liberalization of the grain market may reduce the significance of domestic quality standards, and it is up to the buyer and seller to decide on quality and price. Although a government may retain the use of standards under particular circumstances e.g. for national food reserves, where quality control will be important.

Regional Requirements

End-usage, and hence standards, may vary from one country to the next. Where standards vary considerably between countries, the movement of grain may be hindered. A country may have a single standard that covers both internal and external grain movement. This may not facilitate the trade of a commodity between countries, particularly if the standard permits a greater degree of defective grains than its potential trading partner.

Regional food security may be impaired if the quality of the commodity is not acceptable by some of the countries in the region. Therefore a quality grading standard, acceptable by all users would be necessary for a commodity stored as part of a regional food security programme.

International Trade

International trade of grain may be possible without a standard, if buyer and seller are aware of each others' requirements. However when country-specific information and requirements are not known, a standard will provide the guide-lines that ensure the maintenance of quality, and safeguard consumers' rights. The need for detailed standards may be lessened if a country adopts the "fair average quality" system of assessing quality and price of grain consignments.

Standard Grading of Grain Quality

It was noted above that there is more than one reason to establish standards. Consumerorientated standards tend to specify the nature of a commodity on a pass or fail basis, particularly in relation to wholesomeness or fitness for consumption. Producer or marketorientated standards tend to grade grain into one of several classes based usually on inherent quality and projected market value.

Fair Average Quality (FAQ)

The selling and buying of produce on a Fair Average Quality (FAQ) basis, as practiced by many national and international marketing agencies, is essentially subjective. Normally, samples from different parts of the available stock of produce offered for sale (which may be scattered on farms or in warehouses throughout the producing area) and submit them under seal to independent assessors (public analysts or the like) for appraisal. After examining the samples by sight, smell, taste and (perhaps) touch, the assessors will eslect those samples which they consider representative of the bulk of the samples, mix them together and reduce the lot to a single reference sample which is declared to represent the Fair Average Quality of the seller's stock. Parts of this sample may be used for certain objective tests, e.g. determination of percentage moisture content, oil content, free fatty acid content, or bulk density, if requested by the sellers, buyers or both. In any event, the main part of the reference sample is retained by an independent agency such as the Grains and Feeds Trade Association (GAFTA) for a specified period, during which any transactions involving the produce should be completed. If there is a dispute over quality the independent agency can be referred to for arbitration, and the reference sample may be used as evidence.

It is important to appreciate that the results of the FAQ assessment relate only to the crop which has been sampled, and only for the period agreed upon between the sellers and buyers. If consecutive FAQ same from the same crop, or FAQ samples from the same growing area in consecutive years, or FAQ samples of the same commodity grown concurrently in different areas are compared objectively significant differences in quality may be revealed. Thus FAQ has a loose definition, and can only be applied when fairly wide variations in quality can be tolerated.

The main advantage of FAQ is, of course, that it enables producers to dispose of most of their crop with the minimum of trouble and expense. At the same time the buyer can expect to gain by paying only a moderate price for the crop, although he does run the risk of having to bear the cost of additional processing is quality is some way short of optimal.

Grain Specification

There is a series of Articles in the Commission Regulations, in which relaxations of these grading standards are noted at individual national level.

The Articles also allow a deduction or premium to be made according to changes in individual characteristics, e.g. increased prices paid for lower mc and grain defects, and for higher specific weights (bulk densities).

In comparison, ISO 7970 allows lighter grain with higher mc higher damaged grain, but lower extraneous matter. 'Wheat' as specified must contain a low amount of non-wheat matter, but tolerates higher levels of grain defects, whereas 'wheat' as graded incorporates an element of processing value for the purchaser by rewarding desirable properties.

Conway et al studied quality/value relationships in milled rice. In a system with wellorganised grain quality inspection, they found that qualities apparent at acceptance were only partially reflected in the wholesale price. More cryptic qualities, manifested as change in 'colour' during storage, formed the greater part of the valuation. It is a measure of the success of the application of the acceptance standard that these qualities did not feature largely in the valuation; also, of the valuers' reliance on, or habituation to, the standard. It shows that the acceptance system should not be relaxed, rather that it may need supplementing to cover the newly-identified qualities.

Sampling, Equipment and Methods

Batches of grain are rarely uniform in quality even when regarded as acceptable. Pests usually occur non-randomly in stored grain. Consequently the only sure way of obtaining complete and accurate information about the grain is to carry out a total examination. This may be possible if the quantity to be examined is small, but is usually neither practical nor economical when a large quantity is involved. The choice is either not to examine the consignment at all or to take samples to obtain some information, acknowledging that anything less than a total examination is bound to affect the accuracy of the results.

Principles of Representative Sampling

The results of sample analyses can be expressed in precise terms. However, precise analytical results may be of little practical value, and may be misleading if the samples are obtained without taking into account the non-random or aggregated distributions of foreign matter, damaged grains, insects, etc.

Certain principles of representative sampling must be observed:

- The cnnsignment should be divided into primary units of equal size or status, which may be sampled. For bagged grain, each bag may be regarded as a primary unit. For bulk grain, the primary unit may be expressed in terms of weight, if the grain is being moved, or volume, when it is static - as in a truck or bin.
- All primary units should have an equal opportunity of being sampled. This is possible only during the construction or dismantling of a stack, the loading or off loading of a truck, or when bulk grain is being moved.
- The method should select, without bias, a representative number of primary units from the consignment.

Working Sample Size

In practice, it is necessary to compromise between what is theoretically attainable and the natural desire to obtain results of analyses as quickly as possible. Providing the associated margins of error are recognised and accepted, it is generally suggested that working samples of between 500 and 1000 grains should be used for the determination of common defects such as insect damage, broken grains and discoloured grains. Equivalent minimum working sample weights are:

Maize (small grain)	200g
Maize (large grain)	250g
Sorghum	25g
Black-eyed cowpeas	150g
Wheat	25g
Bulrush millet	10g
Paddy	15g

Samples of these sizes can be analysed in 10 to 20 minutes, depending upon the skill of the inspector and available equipment.

Table 4. Selection of Bags for Sampling.

Number of bags in consignment	*Number of bags to be sampled*
Up to 10	Every bag
11 to 100	10, drawn at random
More than 100	Square root (approximately) of the total number of bags drawn at random according to a suitable scheme.

In the USA, minimum working sample weights of 250g and 1000g are required for the determination of ergot and garlic respectively in wheat, while 250g samples are recommended for the determination of smut in both wheat and sorghum. Such 10 to 40 fold increases on the basic working sample weight illustrate what is meant by 'large' and 'small' sample sizes.

There is a need for sampling awareness when dealing with grain contaminated with mycotoxins. As analytical techniques improve, so detectable and tolerance levels are being lowered. To emphasise the link between the standard test method and specification, Jewers et al. (1989) conclude, "When aflatoxin levels are controlled by legislation it is important that sampling procedures and sample sizes are specified."

For the determination of foreign matter and live infestation, samples should be as large as possible. If bagged grain is being tested, the best results are obtained by passing the entire contents of sample bags over a suitable sieve.

Equipment for Obtaining Primary Samples from Bagged Grain

(i) Simple bag sampling spears

These are the most commonly-used instruments for taking samples from bags, being relatively cheap, simple and quick. Generally, sampling spears having a maximum external diameter of about 12mm are designed for small grains such as wheat, while 25mm diameter spears are suitable for larger grains. To obtain a good cross-sectional sample the spear should be 40 to 45cm in length.

The tapered type of sampling spear penetrates bags easily. However, it takes unequal portions of grain from along the line of penetration, which could lead to distorted assessments of grain quality. More even sampling is achieved with the cylindrical type of sampling spear.

The main disadvantage of obtaining samples with these instruments is that it does not conform to the basic principles of representative sampling. If foreign matter or defective grain happens to be very unevenly distributed in the bag, the haphazard nature of spear sampling could lead to a distorted quality assessment.

(ii) Double-tube sampling spears

These spears comprise two metal tubes, one fitting closely inside the other and each with several common slots. Spears may vary in length from 45cm to 3.5m, and in width from 12mm to 50mm. Turning the inner tube through 180° opens or closes the intake apertures, and so collects grain from a transverse section of the bag.

Double-tube sampling spears are designed primarily for obtaining samples from vertical lines of penetration in bulk grain, although small versions may be used for sampling bagged grain. They are superior in many ways to the simple bag sampling spear, but are still instruments of haphazard rather than representative sampling.

(iii) The Produce-Flow sampler

This sampler was designed at the Tropical Products Institute, now a part of NRI, as a representative sampling device for bagged grain. Grain is tipped into the hopper and falls through onto a cone, which is positioned to ensure that the flow is evenly distributed. Some of the grain is trapped by four vents arranged equidistantly around the base of the cone, and directed via a separate spout into a sample collector. The size of the sample depends upon the dimensions of the vents, which are interchangeable. Sampling of a 100kg bag of grain is complete within 20 seconds of starting the flow.

Equipment for Obtaining Primary Samples from Bulk Grain

Bulk grain is sampled either when it is static, i.e. when it is contained in a truck, barge or storage bin, or when it is on the move, i.e. when it is being discharged through a spout or on a conveyor belt. A wide range of sampling equipment has been developed to meet the special requirements of these various situations, some for small-scale operations and other items for situations where grain is handled in very large quantities.

(i) Equipment for sampling static bulk grain

Double-tube sampling spears

Spears 1.8m long and 3.5cm outer diameter are commonly used, but longer 3.7m double-tube spears are available for sampling grain in exceptionally deep trucks and barges.

The sampling spear should always be inserted into bulk grain at a slight angle from the vertical, with the slots facing upward. The slots must be opened only when the spear has reached the sampling position, and must be closed before it is removed.

Manually-operated deep bin probes

The simplest probe of this type consists of a hollow spear head, which serves as a sample cup, with a spring-loaded cap attached to a metal or wooden rod about 1 metre long. Extension rods are attached to increase the depth of penetration. When the sampling point has been reached a slight upward pull on the rod lifts the cap of the spear head, allowing grain to fill the cup. The probe is then withdrawn completely and the sample removed. A single probe yields up to 300g of sample material.

The deep bin fin-probe consists of a double-tube sampler with a set of extension rods. When the sampling position is reached a twist of the extension rod opens the sample intakes. This action is facilitated by the fin which prevents the outer tube from turning. A reverse twist closes the sample intakes before the probe is withdrawn from the grain. Up to 600g of sample representing a 1.5m long vertical 'cut' may be obtained.

A considerable amount of physical effort is required to push any of these probes into grain. None can be expected to penetrate more than approximately 5 metres.

Pneumatic grain samplers

Pneumatic grain samplers overcome the main disadvantages of manual operation by using powered-suction to penetrate the static bulk of grain, and by taking a continuous sample. They are quicker to operate than manual samplers, and can be used easily to obtain samples from the sides and floors of bulk grain containers.

Auger-type sampler

The sampler consists of a tube approximately 1.4m long and 5cm wide, open at the bottom end and housing a motor-powered auger. Grain lifted by the screw is collected in a bag at the outlet spout. It is necessary to insert the device into the grain at an angle in order to obtain sample material. There are no extension pieces which would permit sampling deeper than the half metre or so the sampler penetrates. The sampler is therefore of limited usefulness.

(ii) Equipment for sampling moving grain.

The Pelican sampler

The Pelican sampler consists of a cowhide pouch attached to a metal frame at the end of a hardwood or tubular metal handle. It is used to obtain samples from freefalling grain, e.g. from a spout discharge to the hold of a ship.

If the spout is sloping, the components of the grain stream are likely to be stratified. It is important, therefore, to cut the sampler through the stream from one side to the other in a single motion to obtain a good sample.

The force behind a stream of grain may be very great. It is essential to observe appropriate safety measures when sampling in this manner.

The Ellis Cup sampler

This is a hand-held scoop , designed for obtaining small samples from bulk grain on moving conveyor belts. When properly used, the cup will obtain a vertical section of the flowing grain at the point where it is inserted into the stream. Samples taken in this way are used for making spot checks on the condition of grain and are not intended as substitutes for representative samples obtained elsewhere in the system.

Sampling with the Ellis cup is hazardous. Extra safety precautions are necessary, as with the Pelican sampler.

Limpet-type sampler

This type of sampler is clamped or bolted to the outside of the delivery spout. A tube is inserted through a hole drilled into the spout wall. The tube usually is open at both ends and has an inlet slot in the upper side projecting into the grain stream. Sampled material is removed either by means of a motorised worm screw, or a plunger operated by compressed air. Worm screw extractors can be made to operate continuously or at intervals. Plunger sample extractors remove samples of fixed size at intervals.

The limpet sampler is capable only of extracting material from part of a grain stream. If there is any appreciable stratification of material in the stream, samples cannot be regarded as representative.

The diverter-type sampler

The diverter-type sampler is probably the best device yet invented for obtaining representative samples from bulk grain. The sampler is designed to take a complete cross-section of a stream of grain, by means of a powered diverter head which takes a cut through the stream, on a preset schedule. During periods of inactivity the aperture of the diverter head is sealed to prevent it collecting dust.

Grain extracted from the main stream by the sampler may be fed directly into a secondary sampler, which reduces the sample to a manageable size before it is delivered via spouting to the grain inspection laboratory.

Quality Determination, Equipment and Methods

The term 'quality' has different meanings for those who are concerned with the handling, storage, processing and utilisation of grain, even though all will be looking for grain of 'good quality'. For example, grain-handling agencies will want dry, insect-free, undamaged grain which will store well; millers will want a grain which will yield a high percentage of finished produce; and consumers will be concerned with flavour, appearance or cooking qualities of grain.

Grain quality may vary with the variety or type of grain selected by the farmer. It will be influenced by climatic and soil conditions during the growing season, cultivation practices, weather conditions at harvest, and by harvesting techniques. Apart from short-term aging or maturation immediately after harvest, quality cannot be improved during storage, handling and processing - on the contrary, it is easily lost.

Every type of grain can be said to possess properties which contribute to its overall quality. A consideration of the various properties or qualities, either alone or together, allows the grain to be graded and valued, and enables the design and development of optimum methods for handling, storing and processing.

Research work into methods for the identification of varieties may hold the key to grain grading systems of the future. Consumers are becoming accustomed to buying fruit and vegetables by varicty, so why not grain? Most cereal crops have been studied; polyacrylamide gel electrophoresis (PAGE) techniques have been used to discriminate between wheat varieties, and reversed-phase high-performance liquid chromatography (RPHPLC) techniques may be applicable to wheat and rice.

Assessment of Grain Quality

With over 420 standard test methods, including at least 75 internationally-applicable, it is apparent that there is a large diversity in grain character. This is obvious when considering the range of uses for grain: paddy to produce milled rice, barley for malting, durum wheat for pasta production etc.

Many assessments are commodity-, product- or end user-specific. Of the wide range of properties, bulk density and foreign matter are commonly assessed for most types of grain. In addition, the influence of moisture content on other grain qualities, as well as the simple economic fact, make it important for quantification.

(i) Bulk Density

All equipment for the determination of bulk density have features of :

(a) causing the sample material to fall from a standard container through a standard height into a standard volume weighing bucket,

(b) levelling the surface of the material in the weighing bucket in such a way as not to influence its packing and

(c) weighing the loaded bucket.

However, differences in equipment design and procedural detail can result in very different values for bulk density, even when the same grain sample is used. It is essential, therefore, that only one type of apparatus is used for determining bulk density. ISO 7971 is a standard reference method with results expressed as mass per hectolitre.

The bulk density of a sample of grain can also be affected by the presence of foreign matter, and varies with mc Consequently it is standard practice to remove as much foreign matter as possible by sieving samples before carrying out bulk density determinations, and also to measure the mc of the sieved material.

(ii) Foreign Matter

Most grain quality standards state that the screens in sieves used for the assessment of foreign matter content should consist of perforated metal plate conforming to specifications laid down by national or international standards organisations. Such specifications cover the composition and thickness of the metal plate, the shape and dimensions of the perforations, and the arrangement of the perforations on the plate.

Operating Capacity of Sieves

The efficiency of a sieve is dependent upon two factors: the dimensions of the apertures in the screen, and the proportional volume of material which will not pass through the apertures. As a general rule, the percentage sieving area' of a screen with small perforations is less than that of a screen with larger holes, and its capacity for sieving efficiently is correspondingly reduced. Also, for a perforated metal screen of fixed specifications the sieving efficiency falls off markedly if the volume of material which will not pass through the apertures exceeds a certain quantity. Table 5 shows the recommended volume of grain that should be placed on a screen, to maintain its sieving efficiency.

Table 5. Grain Sieves, 200mm Diameter, Maximum Loadings.

Nominal aperture mm	*Recommended volume of load cm^3*	*Typical grain equivalent*
8.0	500	300g Maize
4.0	350	250g Sorghum
2.0	200	150g Wheat
1.0	140	100g Millet

Source: International Standard ISO 2591-1973

(iii) Moisture Content

The standard test method (ISO 712) for the determination of mc in cereals is by mass loss in a hot-air oven. The method is time-consuming and a variety of rapid methods have been developed for day-to-day use. These range through accelerated heating by infra-red source gravimetric tests to almost instantaneous readout by electronic moisture meter. Of the latter, two types are common; resistance and capacitance meters.

It is recommended that grain-handling agencies avoid using a mixture of meter types, because this can lead to conflicting results. Instead, the meter best suited to their particular requirements should be selected. The following factors should be considered when selecting a meter to determine moisture content:

1. *Resolution*—the ability of the meter to differentiate between moisture contents which are very close in value. Some meters have the ends of the scale compressed i.e. the scale is not linear. The resolution of the meter is therefore relatively poor for high and low readings.
2. *Repeatability*—a measure of the meter's ability to give a constant reading when the same sample is tested several times. Capacitance meters, due to variations in grain packing, may not produce such accurate results as resistance meters, which normally use a more homogeneous ground or compressed sample.

3. *Reliability*—A measure of variation between meters when measuring the moisture content of the same sample. Meters should be regularly checked and calibrated to ensure reliability.
4. *Stability/drift of measurements*—Affects the frequency of the need to calibrate the meter against the standard test method.
5. *Range of commodity*—Calibrations will be necessary for all the commodities of interest, and the meter must be capable of accommodating them.
6. *Range of mc*—In general, resistance meters cannot measure low mc i.e. lower than approximately 9%, whereas capacitance meters can - to 1 or 2% in some cases.
7. *Sample size*—Meters use differing size of test samples: larger samples give more accurate results, and require fewer replications.
8. *Sample weighing*—Most capacitance meters require the sample to be weighed, thus introducing an extra variable (and extra cost).
9. *Ambient effect*—Meter readings vary with temperature, and correction is required. Some meters automatically display the corrected moisture content.

Role of Standards in Local Trade

Producers of grain will have a number of uses for their produce; as food for the family or livestock, for seed, or for sale. The trade of grains can cover a large area entailing a variety of end usages. A number of end-users would benefit from the uniformity of supply associated with the use of grain standards, such as the commercial producers of food products e.g. beverages, baked products and animal feed, or the procurers for food security reserves. It is difficult to judge whether the setting of standards for these varied uses can best be carried out by the relevant industry rather than a government regulatory body, or a combination of the two. Whatever method is used to formulate the standards, it gives a clear message of quality requirements to producers, and it should provide a more uniform and regular supply for the end users.

Given that the formulation of standards is the best system for providing information to grain producers and end-users, they should be based on factors which both consider important and which are easily recognisable and unambiguous. The selection of grades must allow clear steps which can be easily differentiated and represent a clear change in value and end-use. Standards should be built on those characters that can be accurately and uniformly measured, and interpreted. To assist this process, terminology must not be difficult to understand.

Above all standards should be sources of information intelligible to all and serving a clear function in the production and utilisation of grain.

References

Araullo E.V., De Padua D.B., & Graham M. (1976). R*ice postharvest technology*. IDRC, Ottawa, Canada. 394pp.

Conway, J.A., et.al. (1992) A study in the determination of quality/value relationships in rice. *NRI Bulletin* 55: Chatham, UK: Natural Resources Institute. 45pp.

Foster, G.H. (1982) *Drying Cereal Grains. In Storage of Cereal Grains and Their Products*. C M Christensen, Ed. St Paul: American Association of Cereal Chemists Inc. 79116.

Jewers, K., qt.al. (1989) Methodological developments in the sampling of foods and feeds for mycotoxin analysis. *Journal of Applied Bacteriology Symposium Supplement:* 105S-116S.

Singh G., Bhardwaj K.C. (1985). *Directory of agricultural Machinery and Manufacturers*. Central Institute of Agricultural Engineering, Nabi Bagh, Berasia Road, Bhopal, India.

5

Demand and Supply of Foodgrains

Institutional intervention of some kind in foodgrain marketing and trade in India has a long history and it has passed through several phases. The strongest intervention began in mid 1960s, which has been very closely associated with changes in agricultural technology. Due to various factors like, succession of bad harvests, massive food shortages and near famine like conditions in some parts of the country, challenges to national integrity due to dependence on food aid, and costly food imports, the then government decided to follow policy of self sufficiency in food. This coincided with the advent of new seeds of high yielding varieties of wheat, which later came to be known as green revolution alongwith similar breakthrough in rice. Adoption of these new seeds involved use of modern inputs and investments on the part of the farmers. This made it necessary to create incentive and favourable price environment for the farmers who adopt the new seeds. To achieve this, two new institutions, namely Agricultural Prices Commission (later on renamed as Commission for Agricultural Costs and Prices) and Food Corporation of India, were created that have dominated India's food administration ever since.

Agricultural Prices Commission (APC) was set up in 1965 to advise the government on a regular basis for evolving a balanced and integrated price structure. While formulating such policy the Commission was required to keep in view (a) the need to provide incentive to the producers for adopting the new technology and maximising production and (b) likely effect of the price policy on cost of living, level of wages and industrial cost structure. The policy has been quite effective in encouraging adoption of new technology in the areas well endowed with irrigation and helped in achieving high growth in production of wheat and rice. This made the situation on foodgrain front comfortable, as a sort of balance between demand and supply was in sight by 1980.

With the easing of strain on foodgrain production it was rightly thought to follow policy that leads to balanced allocation of resources towards various enterprises. Thus terms of reference of APC were changed in 1980 to shift emphasis from maximising production to developing a production pattern consistent with the overall need of the economy. The Commission was also renamed as Commission for Agricultural Costs and Prices mainly to satisfy the demand from farmers groups that the Commission fully accounts for cost of production while making recommendation on support/ procurement price.

Commodities covered under minimum support price (MSP) and thus having the price guarantee are: paddy (rice), wheat, sorghum, pearl millet, finger millet, maize, ragi, barley, gram, pigeon pea, moong, urad, tur, rapeseed/ mustard, toria, groundnut, sunflower, soybeen, sesamum, nigerseed, cotton, jute, copra and tobacco while sugarcane is covered under statutory minimum price (SMP). It is illegal for anybody to purchase the commodity at less than its minimum support price when the commodity is covered under the system of statutory minimum price. Apart from major commodities, support price has been extended to some commodities like onion, ginger, potato, castor seed, and some fruits in some states under market intervention scheme.

Here it is pertinent to mention that mere announcement of MSP does not automatically guarantee that the price would not fall below MSP. According to various *Reports of the Commission on Agriculture Costs and Prices*, there are instances of market prices ruling below the support price in some markets for some crops. This shows that mere announcement of support price does not serve the purpose of checking price level falling below the floor if it is not backed by proper market intervention. Experience shows that institutional intervention in ensuring guaranteed price is effective only in those regions and crops where government or public sector agencies procure the concerned commodities in a big way. For instance, official agencies procure wheat and paddy in Punjab and Haryana in a big way, and if price of maize or sunflower in these states fall below the minimum support price there is hardly any intervention by the official procurement agencies. Similarly, if wheat price in markets in say West Bengal or Bihar falls below MSP there is no intervention by official agencies. The purpose of these illustrations is to bring home the point that MSP without effective procurement or intervention apparatus does not guarantee that prices would not fall below the floor set by the government. Implementation of MSP shows that rice and wheat are the main beneficiaries from the policy while cotton at large and edible oil and pulses in some pockets have also benefited from the policy.

During recent years there have been frequent reports from states of Orissa, Madhya Pradesh and Bihar about distress sale of rice and maize below MSP. These states have been late adopter of new technology. Though these states are food deficit at aggregate level, several growth pockets having surplus foodgrains have emerged in these states. These pockets are in the first stage of green revolution and agricultural development, when private trade and market institutions are not in place to provide incentive to encourage adoption of new technology and hence output growth. Agriculture growth would get a serious setback in such areas if institutional support in the form of guaranteed prices is not provided.

One of the serious criticisms of price support policy has been that it has favoured mainly rice and wheat and even in these crops it has favoured the regions which were early adopters of new technology. There is a need to discuss how MSP can be made effective in various commodities and in major producing regions. As it is not feasible to ensure that prices would not fall below MSP in any commodity, can we think of criteria as to what crops should be covered under MSP. Some scholars have suggested that crops which can be considered as price leader or the crops for which technological breakthrough is imminent ought to be covered under the MSP and other candidates for support price would be the crops grown in high risk environment. Vyas further adds that in all these cases MSP should be treated as a transient measure *i.e.* till we are able to have viable crop insurance and or forward trade programme.

Due to changes taking place in consumption basket of food, there is lot of emphasis to develop technologies that promote diversification of agriculture sector. There is thus a need to plan price interventions that encourage agricultural diversification.

There is also a need to discuss criteria on which MSP should be based in the changing context. The popular perception is that MSP is determined based on cost of production. When the emphasis of production is shifting from food security to market led production, is it justified to base MSP on cost of production. Similarly, there are concerns relating to definition of cost of production on which MSP should be based. Some of the cost concepts like Cost C3 are such that the price based on those represent "Profitable Price" rather than "minimum support price".

Government has been seriously influencing structure conduct and performance of rice and wheat markets in various ways like procurement by public agencies, maintenance of stock and distribution etc. To ensure implementation of guaranteed price and to stabilise prices and to run public distribution system, government procures large quantities of foodgrains through FCI and other official agencies from market at a procurement price, which is invariably same as the minimum support price. This blurring between MSP and procurement price has come under severe criticism. It is argued that in order to procure required quantities for PDS and buffer stock, such market conditions are created wherein prices are artificially forced down to the level of procurement prices by measures like putting stock limit, denial of credit, not supplying railway wagons and restrictions on movement of commodities. The second consequence of this is that government is forced to buy whatever produce comes in the market irrespective of its requirements. Thus government has to carry excessive stock which is again sold back after some time for free sale in the market. In this process government has to bear the losses due to quantity and quality deterioration, and inefficient handling and transportation of the produce by the official agencies. It is thus argued that procurement price and MSP should be different. Under this kind of dispensation, government should announce MSP which will provide for only variable cost plus some margin and protect farmer against seasonal price slumps due to gluts. The procurement of quantity required by government should be done at open market price determined by supply and demand in a distortion free market environment. This would have the advantage of buying only the needed quantity, whereas, in the present policy government has been buying whatever is offered for sale.

The foodgrains procured by public agencies are sold either through PDS or at open market price. This system has attracted lot of criticism in the recent years mainly on the ground of efficiency and heavy losses incurred by the public agencies. The government's procurement, distribution, and buffer stocking programmes are reported to have had negative impact of repressing private foodgrain marketing, undercutting its potential contribution to long term food security. This is further said to discourage modernisation of marketing resulting into losses and inefficiencies. The World Bank study proposes that government should use regulatory mechanism only when price movements are outside the desired price band representing width between the ceiling and floor price, which permits reasonable marketing margin for profitable public sector operations. Expenditure Reforms Commission (ERC) set up by the government also recommends that state governments and private trade should be encouraged to enter procurement, trade and export of foodgrains through an assurance of continuity of policy over the next 15 years.

How this shift from public sector dominance in foodgrain trade to private sector would affect farm level prices and price stability? What is the appropriate price band beyond which government should intervene into the market? What are the implications of this band for consumers? These are some of the questions which need to be discussed to understand the implications of proposed changes in government intervention in grain trade.

According to the World Bank study there is considerable scope to reduce the price spread by modernisation of storage, handling, processing and other processes involved in foodgrain marketing. Two major items of price spread are statutory charges and transport cost. It has been observed that that the proportion of produce sold through regulated markets is on a decline and there is growing tendency to sell produce through informal markets to avoid market charges and taxes in regulated markets. This is also happening because of declining credibility of regulated markets to provide competitive price to producer sellers.

There were several formal as well as informal restrictions on inter state movement of agricultural produce, stocking and trading till recently. Even when all requirements are fulfilled there is harassment by official machinery, and one cannot get away without making payment at several places. The consequences of this are: slow movement of produce from surplus to deficit markets, low market integration, depressed prices in producing areas and high prices in consuming areas.

There is system of levy under which millers are required to sell a part of rice milled by them to the government at a price derived from procurement price. The levy on rice is as high as 75 percent in agriculturally progressive northern states. Millers often complain that after contributing as high as 3/4th of rice, at a price which is often below open market price, they are left with small produce to run their business. In reality, the levy component of rice is an important source of economic cost, food subsidy and inefficiency of Food Corporation of India in the rice marketing. What the millers do is that they retain best grade rice with them and supply inferior, broken, adulterated rice to FCI which would sell at a very low price in open market. Except at a subsidised price under PDS, such stock would not be lifted by states for their consumers. Same is the case of custom milling of paddy.

Country has been following a policy of maintaining buffer stock to meet the PDS requirement and for price stabilisation in the wake of year to year fluctuations in production. Maintenance of buffer stocks involves heavy cost on exchequer and suggestions have been made to explore other alternatives for price stabilisation and food security. Some scholars find that option of variable levies turn out to be far superior compared to buffer stock in stabilising prices under liberalised trade, whereas, some other studies find that imports turn out to be much costlier than what appears from international price when country of India's size go for importing food commodity to meet its deficit.

As per the recommendation of ERC a national food security buffer stock of 10 million tonne comprising 4 million tonne of wheat and 6 million tonne of rice should be maintained at all times. The Commission further recommends that objective of procurement policy should be to maintain food security buffer of 10 million tonne and availability of 21 million per annum for distribution through PDS. This way, total buffer stock should not be more than 21 million tonne.

Such restrictions on buffer stock and proposal to reduce role of FCI in procurement for PDS might affect enforcement of MSP. In such situations where should FCI concentrate its operations? Should it continue to procure needed quantity from the traditional regions from where it has been buying earlier or it should shift focus to newly emerging growth pockets.

Several agencies in public sector are involved in the procurement, storage and distribution of foodgrains. Prominent among them are Food Corporation of India (FCI), Warehousing Corporations and state level agencies. Among all these agencies Food Corporation of India has remained in the centre stage of government intervention in foodgrain marketing due to scale of its operation and due to its role in food security. Though this agency has played significant role in ensuring guaranteed price, and hence in adoption of improved technology in traditional green revolution region, its cost of operation and efficiency have remained subject of criticism and are seen as the main factor for mounting food subsidy in the country. Some studies find that FCI cannot be blamed for high cost of foodgrain handling and distribution and economic cost of its operation turns out to be lower than the private trade if latter pays same statutory charges and serve the same purpose which former has been doing. Similarly, Madhura Swaminathan also finds that costs of FCI were justified and that this organisation is vital for food security of the country. In contrast to this, some other studies conclude that there are several negative effects of government of India's foodgrain marketing policy and operations of FCI which are found to be increasingly costly and inefficient. Technical and managerial inefficiencies in FCI operations are said to be responsible for its high costs.

Government has also been intervening directly as well as indirectly in external trade in foodgrains. Though quantitative restrictions (QRs) like quotas and licenses or canalisation through some trading organisation have been relaxed to meet WTO commitments there is still considerable government influence on import and export of foodgrains. Through its procurement policy and stock operations government exercises considerable control over the price at which rice and wheat are available for export. Similarly, imports of foodgrains are subjected to certain stipulations.

The present system of intervention in foodgrains market has come under serious question in recent years. It is felt that the cost of such intervention as reflected in food subsidy bill is mounting year after year and is causing severe strain on country's fiscal resources. There are reports of leakages in the PDS due to which the benefits accruing to target beneficiaries are said to be far from the cost involved. Concern is also expressed about the efficiency as well as desirability of scale of procurement, buffer stocks and distribution system of foodgrains by public agencies in the fast liberalising economic world.

There is a strong feeling that in the new economic environment where buying and selling from international market would be much easier compared to the past the country need not procure large quantity and hold large buffer stock of foodgrains to meet domestic needs (Krishnaswamy 1994). It has also been demonstrated by some studies that trade is an important alternative for price stabilisation than the present policy of buffer stock. Suggestions are also made for increased role for private trade to bring in efficiency in foodgrain marketing system. The biggest criticism of government intervention in grain markets is that it has resulted into piling up of huge stock of grains with government which are unwarranted and are causing huge losses to state exchequer. This has put the country into a paradoxical situation where more than 60 million tonne grains are lying in government stock but one fourth of country's population is not able to have access to even minimum required food.

Though scattered information about various aspects of government intervention in food markets is available in pieces no study in recent years has taken comprehensive view of government intervention in food management that encompasses producers interest, consumers interest, food security, cost to state exchequer etc. In the absence of this, ad hoc and subjective methods have often employed in procurement, buffer stocking and in issuing/releasing stocks.

In the light of this and in the light of different alternatives, which are now available to meet objectives of food policy, it has become imperative to explore more efficient ways of intervention in foodgrain markets.

Indian agriculture in the beginning of 21st century is facing some serious contradictions and challenges. On one hand the country has 60 million tonne of foodgrains in public stock and on the other hand every fourth Indian is reported to be underfed and does not get even minimum calorie required by human body. Similarly, the country is facing massive shortage of pulses and edible oils, which is met through imports, having detrimental effect on the producers in the unfavourable dryland areas, whereas, surplus of wheat and rice is being sold as export at a great cost to the state exchequer. These imbalances are causing serious concern to the country.

The reason for emergence of demand and supply imbalances in Indian agriculture lies mainly in technological and policy factors. The technological breakthrough of green revolution category has been highly biased towards cereals, particularly wheat and rice, which raised their relative profitability and has led to large shift in land and other resources away from other crops. The technological advantage has been further provided strong policy support through remunerative and assured prices for select cereals. This way, rice and wheat enjoy not only productivity advantage but also enjoy stable and assured economic environment. This combination of technological breakthrough and strong policy support has served the important purpose of ushering in a sort of food security in the country. However, the stage has been reached where current level of rice and wheat production cannot be absorbed at existing level of their prices while there is massive shortage of edible oils and pulses. How can these imbalances be corrected? What are the policy options to change the existing production pattern in the desired direction? What kind of input - output pricing policy related to subsidies and price intervention is required to restore balance in structure of agricultural production? These are some of the challenges requiring urgent policy attention. Other challenges confronting policy makers in the area of food policy are: (a) providing growth impetus to agriculture in hitherto underdeveloped regions (b) availability of food to weaker sections of the society at affordable prices (c) preparing farm sector to face competition from trade liberalisations and (d) maintain price stability and ensure remunerative prices to producers.

Food Demand and Requirements

Precise information on future food demand trends and requirement is essential for formulation of food policy. There is lot of confusion about future requirement of food in India, which stems from following reasons. One, the estimates on demand projections by various experts show large divergence. Two, demands for food is often considered synonymous with demand for foodgrains or even cereals. Sometimes very narrow view of food security is taken by looking at availability of only rice and wheat, which creates an erroneous impression about food security. Thus, when the concept of food is taken in proper and broad sense, which includes fruits, vegetables, animal products etc., it may give entirely different demand scenario compared to the one which equates food to grains alone.

Demand projections

Two sets of estimates giving demand projections for food towards the year 2020 have been widely discussed in the recent years. One set is based on the study by Bhalla, Hazell and Kerr published by Washington based International Food Policy Research Institute (IFPRI) and the other one is based on the study by Kumar published by Indian Agricultural Research Institute, New Delhi. The two sets basically differ on two counts viz. estimates of expenditure elasticity, which is at the core of demand projection, and estimates of feed demand.

IFPRI study is designed to include changes in elasticity parameter over time, whereas, study by Kumar assumes same elasticity throughout. There is wide divergence in elasticity estimates used by the two studies in making demand projection. For example, assuming 3 percent growth rate in per capita income during the period 2000 AD to 2020 AD, IFPRI study would imply about 8 percent growth in per person cereal demand whereas Kumar's study implies 1.6 per cent decline in per capita cereal demand. Thus, according to Kumar, direct demand for cereal would rise only on account of increase in population, whereas, Bhalla, Hazell and Kerr (hereafter referred as BHK) report large increase in demand due to growth in income beside the impact of population growth.

According to Kumar any increase in per capita income of Indian population would result in a small decline in direct consumption of cereals and modest to high growth in demand for livestock products, fruits and vegetables. BHK finds that income growth would raise demand for livestock products at a much higher rate than that reported by Kumar and they are not in agreement with Kumar on negative income elasticity of cereal demand. However, scholars working in this area like Murthy (1998, 1999) and Rao also observed decline in per capita foodgrain consumption in rural India in the recent period which lends support to negative, albeit small, expenditure elasticity of cereals reported by Kumar.

The second source of divergence between the projections by Kumar, and Bhalla, Hazell and Kerr is the estimate for feed demand. BHK assumes that expansion in livestock production would entail increase in feed coefficient to level of 1.2 kg of cereal per kg of meat and egg and 0.72 kg of cereal per kg of milk as they observed that traditional sources of feed like grazing areas are shrinking. They also assumed higher growth in demand for livestock products compared to Kumar, which further increases the difference in feed demand estimates of the two studies.

According to BHK, domestic demand for cereals would grow at the trend rate of 2.53 per cent corresponding to 3.5 per cent annual growth rate in per capita income. According to Kumar the rate of growth in cereal demand would be 1.88 per cent.

Now the most important question is which growth rates is likely to hold for India. The two estimates have been widely discussed in various seminars where the consensus is that actual demand growth would be within these two estimates as the higher side estimate is believed to be over-estimate and the lower one is believed to under-estimate the cereal demand. The average of the two comes to 2.20 per cent, which can be taken as reasonable estimate to reflect growth in cereal demand in India.

As against the growth rate of demand, Kumar has projected somewhat higher growth in supply of cereals, which would leave positive surplus of total cereals towards the year 2020 even under deceleration in total factor productivity growth. Based on Kumar's projections India is not going to face deficit of cereals up to 2020. Further, Kumar has noted significant changes in composition of food basket which has implications for domestic production and resources allocated to various enterprises. Demand for non cereal foods is projected to grow by more than 50 per cent the growth in cereal demand. Demand for pulses, edible oils and vegetables would increase in the range of 2.9-3.0 per cent and that of fruits and livestock product by more than 3.20 percent annually. Growth rates achieved in supply in the recent past are higher than the growth rate in demand, for all the commodities except pulses and oilseeds. In case the growth in supply of deficit commodities fails to keep pace with the trend in demand, the gap has to be filled either through imports or it would result in increase in relative prices of the concerned commodities.

Demand Driven Crop Diversification

Growth in supply of edible oil and pulses has not been keeping pace with the growth in demand. This has resulted in fast growth in import of edible oils while pulse deficit is reflected in both imports as well as in the incrcase in domestic prices of pulses. In the beginning of 1990s India imported very small quality of edible oil which further declined to around 100 thousand tonne by 1993-94. After this year, edible oil imports witnessed sharp increase. During 1999-2000 India has emerged as the largest importer of edible oil in the world with import exceeding 4 million tonne involving total value of about Rs. 8000 crore. With such magnitude of import, India meets more than 40 per cent of its edible oil demand from other countries.

Sharp increase in edible oil imports resulting from liberalisation policy has depressed domestic prices considerably. Though palm oil constitutes predominant share in edible oil the impact of import is realised on all edible oilseeds due to close substitution among different edible oils. During 1993-94 to 2000-01 price of edible oil in rupee has increased merely by 3 per cent whereas prices of rice, wheat and pulses in the same period have increased by 68, 78 and 79 percent respectively. Thus, relative prices and profitability of edible oils has sharply declined during the last 8 years causing adverse impact on farmers growing oilseed crops.

Like edible oils, India depends heavily on import to meet its pulse's requirement. Average import of pulses during 1998-99 to 2000-01 was 416 thousand tonne, which constituted 3.4 per cent of domestic demand for pulses. During the year 2001-02 import of pulses exceeded 2.1 million tonne which corresponds to about 16 percent of domestic demand for pulses. Despite reliance on imports, pulse deficiency in Indian diet has been increasing continuously. During the year 2000 per capita per day net availability of pulses in India dropped to 31.2 gm, which is less than half of the level that existed during early 1960s i.e. before the onset of green revolution in India.

Compared to massive shortages of edible oils and pulses, India is facing serious problem of selling its cereal surplus. Buffer stock of rice and wheat before arrival of wheat crop of rabi 2000-2001 was reported to be 46 million tonne and after the procurement of rabi season of the year 2001 the buffer stock with central agencies has exceeded 60 million tonne. During the agricultural year 2000-01 more than one fourth of rice and wheat output of the country has remained in public sector stocks. This is causing serious strain on state exchequer due to cost of storage, interest on blocked capital and deterioration in value of stored produce.

How can these imbalances be addressed? What are the hindrances and constrains to diversify some of the area from rice and wheat or other cereals towards pulses and oilseeds to achieve balances in domestic needs and production? We begin to look at answers to these questions by examining the differences in net return to farmers over paid- out cost for various crops. These estimates of net return refer to triennium average ending 1996-97 far kharif crops and 1997-98 for the rabi crops. This is the latest triennium for which published data is available from 'Cost of Cultivation Scheme' of Directorate of Economics and Statistics, Ministry of Agriculture, GOI – the only set of data which is comprehensive, comparable and representative for the major states producing the referred commodities.

Among kharif crops cotton and in rabi season wheat are found to be the most remunerative in most of the states. In Andhra Pradesh, urad was highest paying pulse crop but its net return was only Rs. 6790 compared to Rs. 10098 from paddy. Similarly, net return from groundnut (oilseed) was quite low compared to paddy. Net income from rapeseed/mustard cultivation in Punjab is less than one fifth of the net income from wheat. In Rajasthan, chickpea requires 130 per cent and rapeseed/mustard needs 72 per cent increase in net income to compete with wheat.

Relatively low return is one of the factors for production of oilseed and pulses not keeping pace with the domestic demand. The other factors are high uncertainty and risk associated with their yields and prices. Prices of rice and wheat in the representative market of the country deviated around trend by less than 9 per cent. Compared to this, instability in prices of pulses, as indicated by chickpea, was 28.75 per cent. Similarly, instability in prices of rapeseed/mustard was more than double the instability in rice and wheat prices.

The above results show that cultivation of pulses and oilseeds in India is characterised by low returns and high degree of yield and price risks. Therefore, diversification towards these crops would require increase in yield and prices along with stability in productivity and prices.

Prospects of Grain Exports

Under liberalised trade, export and import are considered the main instrument for stabilisation of domestic demand and supply. However, due to high volatility in international prices and because of being residual exporter, India could not use trade options very satisfactorily to stabilise supply of foodgrains. The trade option has particularly become difficult to dispose off large stock of rice and wheat in India in international market in the recent years.

Despite surplus stock, export of rice (non basmati type) witnessed a big setback during 1999-00 and 2000-01 mainly because of two reasons. One, international prices of rice have dropped sharply even in Rupee terms after 1998-99. Two, domestic prices have been moving up mainly under the pressure of increasing support prices. This has reduced competitiveness and profitability of rice exports. The net result has been piling up of rice in government stocks which has remained above 22 million tonne during early months of the year 2001 as against 15 million tonne a year back. In order to reduce the level of buffer stock to reasonable level, Food Corporation of India had to sell rice at Rs. 5650 and parboiled rice at Rs. 6000 per tonne to traders for export even though open market price in the country is ruling around Rs. 9500 per tonne during the year 2001. This offer has attracted good response as private traders are able to book export orders at a price of $ 175 to $ 210 or approximately Rs. 8242 to Rs. 9891. However, these exports are causing huge loss to FCI because cost of acquisition of rice to FCI works out to be around Rs. 10230 per tonne. Corresponding to these costs and export price, the country is incurring a loss of more than Rs. 4290 for each tonne of export of non basmati rice. The estimate of loss is worked out as under:

1. Support price of paddy: Rs. 510 for common grade and Rs. 540 for grade A
2. Ratio of cost of acquisition of one quintal rice to MSP of paddy: 1.98 (Annual report of Ministry of Consumer affairs and PDS)
3. Cost of acquisition of 100 kg rice: Rs. 994 common grade and Rs. 1052 grade A
4. Price charged for stock released for export: Rs. 565-600/ quintal
5. Difference between (3) and (4): Rs. 429 to 452 / quintal

Actual loss would be much higher if cost of storage and transport are fully reckoned.

As has been the case with rice, international prices of wheat have experienced sharp fall during the recent years. Despite a decline in the $ - Rupee exchange rate, international wheat prices after 1996-97 declined in Rupees term by 20, 14 and 7 percent in the next three years. In contrast to this, domestic prices of wheat registered upward growth, mainly on account of raise in support price, rendering import of wheat profitable. As a consequence, 1.333 million tonne wheat was imported by the private traders during 1999-00 despite record level of buffer stock in the country.

The more recent experience of wheat export (2001-02), resulting from the bid to dispose off excessive stock with the government, is yet another evidence of India's failure to get international price for wheat. In contrast to international price of around $130/tonne, India found it difficult to sell wheat abroad even at a price of $103 per tonne. As against economic cost of Rs. 8300/tonne to Food Corporation of India, and open market price of Rs. 7000/tonne, wheat was offered for export at Rs. 4300 per tonne for May

2001. This amounts to implicit subsidy or loss of Rs.4000 per tonne of wheat export.

The justification given for export of rice and wheat at a huge loss and at a price below domestic market price is that there is no alternative to dispose off the massive stock of wheat that has accumulated with government agencies. Releasing such stock in open market, rather than its sale as export, would put downward pressure on domestic prices and might force domestic prices to go below minimum support price. The more serious implication of such move would be its adverse impact on future output growth and supply. On the other hand, demand side factors favour reduction in domestic price on two counts. One, on the ground of market clearance and two, to improve access to food as more than one fourth of the population of India can't afford to buy adequate food at existing prices. Thus, dealing with the situation of wheat surplus is posing challenge to the government.

The government needs to take different kinds of measures to address the immediate problem of surplus and to sustain food security in the long run. In the short run, surplus stock of grains should be diverted for food for work and alike employment generation programmes, at a price not higher than what is fetched from export. The real food subsidy for such uses is represented by the difference between the opportunity cost of stock, i.e. export price, and the price at which the produce is offered under the programme. This would hardly involve any extra cost to the state exchequer. In the medium run, it would be desirable to bring balance between demand and supply factors. This would involve a check or even small decline in minimum support price in case international price of wheat remains low or do not provide scope for export at remunerative price. The long run policy should aim at adjustment in crop pattern consistent mainly with domestic requirements. Price policy in the long run should ensure increased participation of private sector in the wheat trade.

Declining Demand for Grains

In India large population is reported to be below poverty line and is undernourished. Cereals are the main source of calorie and food security for such population. In the recent years huge quantity of cereals (rice and wheat) has piled up in government stocks which has reached a level of about 60 million tonne. This constitutes more than one third of the total production of rice and wheat in the country and is far above the quantity genuinely needed in the buffer stock in the country.

Some analysts believe that this has reduced availability of grain to needy public and is being attributed to various factors like high rise in procurement prices of rice and wheat after 1990-91. To find the exact nature of change in cereal demand and its causes we have looked into the trends in per capita production and availability of cereals, prices of various commodities and changes in consumption pattern in the country.

For the sake of proper correspondence between production and consumption and agricultural year and financial year, production of rabi and kharif in the same calendar year are added together. For example, 1971 includes output of Kharif season of 1971-72 and rabi output of the year 1970-71. Rabi output of cereals shown under the year 1970-71 as per the official documents, include output of rabi crops whose harvest normally starts in March and extends to the month of May. Kharif cereal output put under year 1971-72 in the official data refer to the output almost all of which is harvested in 1971. Thus, to study the relationship between different variables reported on the basis of financial year (April – March) on the one hand and cereal output on the other hand, it seems reasonable to use cereal output harvested during March to December of the same year rather taking output harvested during July to next year June as followed in the official statistics.

Past trends in cereal production and availability show that both, per capita production as well as availability of cereals, have moved on a somewhat rising trend during 1970 and 1980s. Per capita production of cereals reached peak level in TE 1990 when India produced 524 gram cereals per capita per day. Since than, cereal production has risen almost at the same rate as the growth rate in India's population. Accordingly, per person production of cereals during 1990s fluctuated around level of 520 gram.

During 1970s and 1980s, trend in per capita availability, which in a somewhat crude sense can be considered to represent per capita consumption, kept pace with trend in per capita production of cereals. However, during 1990s per capita consumption in cereals shows signs of decline. This way, the decade of 1990s extending to earlier year of first decade of new Century, shows two clear indications relating to cereal production and consumption in India. One, growth rate in cereal production decelerated to the level of population growth after 1990. Two, average availability of cereals has followed a decline in the recent years and it has gone down to the same level as observed during 1987 which was a very bad drought year.

However, the recent decline in per capita availability is not a result of slack in production nor it has resulted from export of cereals. In fact the decline in per capita availability of cereals despite no similar decline in production is explained by accretion to stock of cereals held by various government agencies. Net addition to stock has witnessed a very large increase in the last four years. Even during the year of very poor production, stock show increase to the tune of 7 percent of total production of cereals.

Another aspect of total cereal use in the country is its composition in terms of open market supply and supply through public distribution system which is sold at a subsidised or concessional rate. The decline experienced in per capita availability of cereals in the recent years resulted from decline in both, open market demand as well as PDS supply. Thus decline in per capita availability of cereals in the country in the recent years has affected all kind of consumers – those depending upon PDS as well as those depending upon open market.

The vital question now is why cereals have gone into stock rather than being consumed by people. There are three possible explanations for this.

Table 1:Per capita availability of cereals in open market, under PDS (in gram)

Year	*Total*	*Open market*	*PDS*
1990-91	449.3	414.8	34.59
1991-92	458.6	416.4	42.28
1992-93	425.8	387.6	38.24
1993-94	437.7	403.0	34.71
1994-95	442.3	412.9	29.38
1995-96	463.7	428.5	35.18
1996-97	465.5	423.1	42.40
1997-98	418.3	381.8	36.48
1998-99	438.6	393.9	44.68
1999-00	422.3	383.4	38.92
2000-01	395.4	367.6	27.73

Source: Economic Survey, various issues.

One, as procurement and issue prices fixed by the government play a major role in determining market price of cereals, sharp rise in these prices during 1990s is said to be the cause for decline in per capita consumption. Two, change in taste and consumption pattern could be the reason for the decline in cereal consumption. Three, decline in quantity supplied through public distribution system can also be the cause for overall decline in cereal consumption.

Price Trends

Changes in consumer level prices can be seen from the changes in rural retail prices of wheat. During the decade of 1980s highest increase in wheat prices was recorded during 1988-89, which followed very bad drought of year 1987, when prices shot up by 20 percent. During 1990s rural retail prices increased by close to 21 percent during 1991-92 and by as high as 41 percent during 1996-97. Similarly, PDS price of wheat was frequently revised upward during 1990s. The biggest increase was witnessed in 1998-99 when PDS price of wheat was raised by 44.44 percent. Because of this raise in PDS price, average retail price of wheat in the country in open market turned lower than the PDS price. PDS price of wheat was again increased by 22 percent in 2000-01 eventhough offtake from PDS was on a decline. In the same year PDS price for population below poverty line was increased by 65 percent. The impact of these increases in PDS prices is reflected in sharp decline in PDS supply which dropped to mere 27 grams/ person/day during 2000-01.

As has been the case with wheat, procurement prices of rice witnessed much higher increase during 1990s as compared with 1980s. During the five years between 1989-90 and 1993-94 rice procurement price was raised by about 11 to 17 percent each year. However, rural retail prices of rice did not show higher increase during 1990s as compared to that during 1980s. In contrast to free market price in rural India, PDS price of rice witnessed hefty increases in most of the years during 1990s. This first narrowed down the difference between open market retail prices and PDS prices and then rendered PDS prices higher than average of rural retail price during 1998-99. Similarly, PDS price for population below poverty line was hiked by 61 percent during the year 2000-01.

Changes in Rice and wheat Prices

It is interesting to analyse behaviour of rice and wheat prices vis a vis prices of other foods relative to overall index of consumer prices in the country. This has been studied by examining the behaviour of rice, wheat and food prices deflated by the general consumer price index (CPI) with base

1982=100. During the entire decade of 1980s procurement prices of wheat increased at a slower rate as compared to food prices and overall CPI. The rate of increase in procurement price of wheat turned out to be faster than CPI and other foods during 1990-91 to 1992-93 which set off the decline faced during 1980s. Again there was a decline in procurement price of wheat relative to prices of other foods and other commodities for three years after which procurement prices in real terms moved closer to the level of early 1980s.

Rural retail prices of wheat in real terms followed decline in early 1980s and than fluctuated sharply. Retail prices of wheat relative to CPI remained lower throughout.

Series for rice prices show that real procurement prices for paddy have declined throughout during 1980s as has been the case with wheat. This was followed by faster growth in procurement price of paddy relative to consumer price index for about five years which turned procurement price favourable in real terms only for two years after which again growth in procurement prices turned lower than general CPI. Rural retail prices of rice, in real terms, followed declining trend during 1980s as well as during 1990s.

Procurement prices of wheat increased by 10.53 percent annually during 1990s which is more than double the growth rate recorded during 1980s. Similarly, procurement price of paddy show annual growth rate of 9.65 percent during 1990s compared to 5.42 percent during the previous decade. Wholesale prices of wheat and rice, which were strongly influenced by government intervention, also showed much higher growth during 1990s compared with 1980s. Second, increase in procurement prices during 1980s was quite smaller compared to the growth rate in prices of food and other commodities, whereas during 1990s, growth rate in procurement prices turn out to be higher than the general rate of inflation.

Growth rate in rural retail prices also showed increase during 1990s but the increase was much smaller compared to the increase in procurement prices and in wholesale prices of rice and wheat. During 1980s growth rate in rural retail prices was higher than the growth rate in wholesale prices of wheat whereas during 1990s retail prices showed lower growth rate compared to wholesale and procurement prices. This shows that high growth rate in procurement and wholesale prices was not backed by demand side, causing the retail prices to trail behind the wholesale prices. This also caused adverse impact on participation of private sector in rice and wheat trade.

DIVERSIFICATION IN CONSUMPTION PATTERN

While recent decline in per capita demand for cereals from PDS seems to have resulted from sharp increases in PDS price for APL consumers, the decline in open market demand seems to be caused by increase in open market retail prices of rice and wheat as well as by non price factors. Beside prices, demand is also affected by change in income and preferences. Both in rural as well as in urban areas per capita consumption of rice, wheat and coarse cereals has declined sharply since 1987-88. In contrast to this, per-capita consumption of edible oil, vegetables, fruits and meat, egg and fish has shown appreciable increase. This provides strong evidence that in India food consumption pattern is getting increasingly diversified towards non-cereal products.

Table 2. Changes in food consumption pattern in rural and urban India, 1977-1999 (quantity in kg/person/annum)

Items	*Rural*				*Urban*			
	1977	*1987*	*1993*	*1999*	*1977*	*1987*	*1993*	*1999*
Rice	86.5	88.1	85.4	81.0	67.6	68.1	64.2	62.5
Wheat	49.4	61.6	53.5	53.9	64.6	60.4	57.4	55.4
Coarse cereals	56.7	29.8	24.1	17.7	14.8	10.6	7.7	7.1
Total cereals	192.6	179.5	163.0	152.6	147.0	139.1	129.3	125.0
Pulses	8.7	11.5	9.2	10.1	11.7	12.2	10.5	12.0
Milk/milk product	24.6	58.0	51.4	50.5	39.7	64.9	68.3	72.4
Edible oils	2.7	4.3	4.6	6	4.8	6.8	6.3	8.6
Vegetables	24.7	50.8	53.2	66	39.7	66.4	63.1	70.0
Fruits	2.6	10.3	9.8	17	5.9	18.8	20.1	19.0
Meat, egg, fish	2.7	3.3	4.1	5.0	4.8	4.9	6.3	6.8
Sugar and gur	13.5	11.0	9.2	10.1	17.1	12.3	11.8	12.0

Source: NSSO

There are scholars who view this change in consumption basket as a healthy trend but some researchers consider it as adverse to consumers. According to former, increasing mechnisation of human activities, particularly in rural areas, has reduced physical exertion and access to modern transport facility is bringing improvement in comforts. All these add up to reduced calorie requirement. Secondly, when income improves, there is a tendency to improve variety in food intake and go for nutritious and healthy food rather than increasing calorie intake. Improvement in communication and transport network and development of marketing system has made food produced in one region easily accessible in other and far flung regions. This has improved supply of non-cereal foods at micro level, and has contributed towards diversification in food consumption.

On the other hand, some scholars consider decline in per capita cereal consumption as adverse for the consumers, particularly for low income group. It is argued that improvement in per capita income should result in fast increase in the indirect demand for cereals which should lead to increase in overall per capita demand for grains in the county. The argument goes like this. Improvement in income leads to fast growth in demand for livestock products, which should result in higher growth in the use of food grain as feed. This arguments has been buttressed by citing example of China and by showing that per capita total food grain use in countries with better income is higher than that in India. Here, there is a need to distinguish Indian situation from other countries. Despite observed growth in demand for livestock products, per capita consumption of meat, egg and fish in India remains very low. This seems to be largely due to cultural factors. Even if it can be afforded, people in India do not eat meat etc. on a regular basis. For a vast majority of those who are non-vegetarian, meat intake is preferred occasionally and not as a part of regular diet. Therefore, India is not likely to witness serious competition between food and feed uses of grain. Similarly, cultural difference between India and other countries need to be kept in mind when comparing per capita food consumption, particularly food grain consumption. According to Indian value system, there is emphasis on eating only that much which is needed for survival and there is no tendency to eat more and expend the same by doing exercise etc. as is the case in developed countries. In the light of these factors, cereal demand in India should not be expected to grow at faster rate than the population growth rate. This has implications for planning agriculture production in future.

Market Arrivals and Procurements

Initial objective of procurement of foodgrains by the government agencies was to ensure remunerative prices to producers and reasonable prices to consumers and to maintain price stability. Since private trade was not performing these tasks satisfactorily, government had to intervene in grain markets. As a result of this intervention, government procured substantial part of marketed surplus from food surplus states. Procurement by government constituted 10 to 15 percent of total rice output during 1985-86 to 1989-90. In the case of wheat 15 to 22 percent of production was purchased by government. When India initiated economic reforms in year 1991, a move towards increased participation by private sector in various spheres of economic activities was started which in other words imply reduced role of Government in marketing and services. Government started gradually reducing its participation in industry and services sector. In the case of agriculture, particularly the food grain trade, some reduction in the role of government took place after 1992-93. Between 1992-93 and 1996-97 percent of wheat output procured by government declined from 22.3 percent to 13.41 percent.

Table 3: Production and procurement of rice and wheat by official agencies, 1985-86 to 2000-01

Year	*Production Lakh tonne*		*Procurement Lakh tonne*		*Procurement as % of production*	
	Rice	*Wheat*	*Rice*	*Wheat*	*Rice*	*Wheat*
1985-86	638	471	99	103	15.51	21.89
1986-87	606	443	92	105	15.19	23.69
1987-88	569	462	69	79	12.13	17.11
1988-89	705	541	77	66	10.92	12.20
1989-90	736	499	118	89	16.04	17.85
1990-91	743	551	127	111	17.09	20.13
1991-92	747	557	102	78	13.66	14.01
1992-93	729	572	130	64	17.84	11.19
1993-94	790	591	143	128	18.10	21.65
1994-95	818	658	137	119	16.75	18.09
1995-96	770	621	100	123	12.99	19.81
1996-97	817	694	130	82	15.90	11.82
1997-98	825	663	155	93	18.78	14.02
1998-99	860	708	126	127	14.65	17.94
1999-00	895	756	182	141	20.34	18.66
2000-01	849	687	208	164	24.50	23.87
2001-02	907	735	202	206	22.27	28.03

Source: Bulletin of Food Statistics, Ministry of Agriculture, various issues

Similarly, government procurement of rice declined from 17 percent in 1993-94 to about 14 percent in 1996-97. However, this decline turned out to be the part of cyclical variation resulting from production fluctuation as experienced in the past and government procurement witnessed steep increase in the recent years. The difference in the current and the past situation is that hike in quantity procurement by government took place despite very comfortable situation of food stock. During the year 2000-01, government had to procure all time high share of wheat and rice produced in the country. This increase took place despite two odds. One, production during year 2000-01 was low compared with the past 2 years and two, buffer stock with government was already double the requirement for such stocks. This shows that in the recent years private trade played much-reduced role in grain trade. The reason for this could be huge build up of bufferstock. Private sector had a feeling that government would be required to reduce the level of excessive stock, which would dampen the domestic prices and might cause losses to the private sector. This way, the recent years witnessed decline in the role of private trade in grain marketing, which is not a healthy trend for long run development of grain trade in the country.

Another serious criticism of government intervention in grain markets is the regional bias in its operation. Official procurement is alleged to be helping a few states and concentrating their operations in already developed states. States like Bihar, Gujarat, Madhya Pradesh, Karnataka and Maharashtra have experienced sharp increase in marketed surplus of wheat during 1990s. Similarly, percent of marketed surplus of rice has witnessed substantial increase in several states like Bihar, Karnataka, Madhya Pradesh, and Orissa. But these states have hardly benefited from government procurement.

Table 4: Market arrivals of wheat as % of production in selected states

State	***TE*** *1983-84*	***TE*** *1986-87*	***TE*** *1990-91*	***TE*** *1993-94*	***TE*** *1996-97*	***TE*** *1999-00*
Bihar	10.87	10.93	15.20	17.13	16.20	17.13
Gujarat	38.10	38.00	45.70	50.47	52.30	63.20
Haryana	37.80	40.93	41.93	46.17	43.80	44.57
Karnataka	9.03	8.17	8.93	16.53	25.43	29.07
M.P.	10.43	10.37	10.77	17.27	24.17	40.50
Maharashtra	26.47	27.40	30.03	44.90	38.60	59.03
Punjab	52.97	47.67	47.27	50.77	54.20	51.33
Rajasthan	19.00	20.67	18.87	19.20	18.63	20.33
U.P.	15.63	17.07	18.60	17.80	17.67	23.40
All India	26.80	27.27	27.87	29.77	30.27	34.17

Source: Bulletin of Food Statistics, Ministry of Agriculture, various issues

It is striking to observe that 88 percent of wheat that arrived in market in Punjab was procured by official agencies whereas only 2.2 per cent of marketed surplus in the states other than Punjab, Haryana, Uttar Pradesh and Rajasthan was procured by official agencies. Similarly, in the case of rice more than three fourths of market arrivals in Punjab, 59 percent inHaryana and 68 percent in Andhra Pradesh was procured by government agencies. Government agencies purchased between 8 to 37 percent of rice brought to market in Madhya Pradesh, Orissa, Tamil Nadu and West Bengal. In other states, government procurement was below 3 per cent of market arrival. This shows very strong regional concentration of government intervention in selected few states.

Information furnished above shows that the main beneficiaries of wheat procurement have been Punjab and Haryana and to some extent UP and Rajasthan. In the case of rice main beneficiaries have been Punjab, Andhra Pradesh, and Haryana while states like Uttar Pradesh, West Bengal, Orissa, Tamil Nadu have also been benefited to some extent. The remaining States have hardly been benefited from government procurement of grains even through these neglected states have shown tremendous increase in generating market surplus.

References

Acharya, S.S.. Agricultural Marketing in India: Some Facts and Emerging Issues, *Indian Journal of Agricultural Economics*, Conference no. 53(3): July – September.

Kumar Praduman. Food Demand and Supply Projections for India, *Agricultural Economics Policy Paper* 98-01, Indian Agricultural Research Institute, New Delhi.

Piggot, R. and D.S. Sidhu. *Efficiency of Foodgrain Marketing in India*, Background Working Paper, World Bank, Mimeo.

Rao C.H. Hanumantha. Declining Demand for Foodgrains in Rural India: Causes and Implications, *Economic and Political Weekly*, Vol. 35 No. 4, Jan 22.

World Bank. *India Foodgrain Marketing Policies: Reforming to Meet Food Security Needs,* vol.I and II, Report no. 18329-IN, April 17.

6

Foodgrains Policy and Management

There has been a popular feeling that the recent inflation is caused by poor foodgrain management. As is so often the case with complex economic matters, the reality is more complicated than popular perception. In the case of the recent Indian experience, it is both true that food inflation is high and our foodgrain management leaves a lot to be desired. But it is not clear that the latter is causing the former. Poor management may keep the prices of some foodgrains higher than they need be, but inflation, defined as a sustained increase in price, is not caused by this, which would typically require a sustained deterioration in foodgrain management for which there is no evidence.

This, however, does not change the fact that we need to improve our agricultural productivity and the management of the market for foodgrains. The latter is the subject matter of this paper. The flaw in the system is that, while we have steadily procured foodgrains, especially wheat and rice, we have not done equally well in releasing the grain when the need arises. Doing the former and not the latter has meant that the net effect has been to raise the average price of food. Also, a good market intervention, as discussed in greater detail in the next sections, entails buying up when prices are low and selling when prices are high. Comparing procurement and inflation data it is clear that little effort has gone into creating such a cycle. The years 2006-07 and 2007-08 were ones of low inflation for both rice and wheat. There is no evidence of extra procurement in those years. In fact, wheat procurement actually fell in those two years. Given that 2009-10 is an inflationary year, one would have expected lower procurement but that was not the case.

The theory of how we can improve our foodgrain management will be discussed in the next sections. However, some simple lessons that will come out of that can be stated in advance. First, we have to have a ready set of rules of how and when to release foodgrain, a kind of Standard Operating Procedure (SOP). There should be no need to have special cabinet committees to take the decision. If prices are rising, there has to be a rule about the *automatic* release of food. Moreover, the release should be in small batches – the reason for this will become clear later.

Second, after we release the food, we should not try to excessively monitor what the buyer of the food does with it. As per present practice, the food that is released through open market operations by the Food Corporation of India (FCI) is sold to millers, and only rarely to traders. These millers are then prohibited from selling the wheat to yet other buyers and make profits from this. However, if our aim in releasing food is to lower the price, it is not clear why we should prohibit further reselling of the food. And the instinctive urge to prevent anybody from making a profit – and creating a bureaucracy to monitor this – does more harm than good. What we should do instead is allow people to strive to make a profit but design a system which uses this striving to achieve social well-being.

The consequence of having such severe rules about what the buyer of food from FCI can do with the food means that the zest for buying FCI food will be low. At times when the price of food is high and we want to lower it, it is a mistake to curb the traders' and the millers' freedoms. This would result in poor offtake of foodgrains and prices remaining high. In general, we manage to release less food than we plan and, moreover, the release is even more inadequate when done through the Open Market Sale Scheme (OMSS).

The right policy is to place as little restriction on the buyers of foodgrains as we can and to permit them to make profits by selling the foodgrain to the ultimate consumers. The profit of the trader and the miller is of course not the aim of the government but it is the instrument through which government can reach food to the poor. This does not mean that we should not have any strategy to limit the profit, but simply that this must not be done by creating disincentives for the trader or the miller for buying up grain and selling it to the ultimate consumer. The secret of keeping profits low and delivering food to the ultimate consumer is to release the procured grain in small quantities to large numbers of traders and millers and giving them the freedom to make profits. Competition will drive prices down through natural market forces. Good policy consists of exploiting the laws of the market, not denying that they exist

Intra and Inter Year Price Fluctuations

Rice and wheat are the staple food for Indian population. Therefore, their prices cause significant impact on the well being of general population. These two commodities being wage goods, determine consumption level of vast majority of population and overall price level of the economy. In the wholesale price index (refers to base 1981-82), agricultural commodities account for 27.45 per cent weight, out of which rice and wheat respectively account for 13.4 per cent and 8.2 per cent weights. In the overall price level, rice and wheat constitute 3.68 per cent and 2.25 per cent weight respectively. Due to their importance in consumption basket and in determination of the overall price level, prices of these two commodities attract considerable attention of the policy makers. Stability in prices of these two commodities has remained an important goal for the country. Because of all these factors there has been considerable interest to know the factors affecting yearly and monthly variation in prices of rice and wheat. One of the obvious factors affecting price level and changes therein is production level of the commodities. However, it has been often observed that even during the years of normal production sometime prices tend to show abnormal rise, which underscores the need to identify the other factors affecting changes in rice and wheat prices. The present exercise is a step in this direction to analyse various factors affecting intra-year and inter-year variation in prices of rice and wheat in the country.

Analytical Procedure

With respect to time, there is two kind of change in prices of agricultural commodities viz. intra-year and inter-year. Sometime intra-year variation may not be captured by year to year variation in prices. This is particularly true in the case of agricultural commodities where there is strong seasonality. In such cases it can so happen that monthly price spread can go very high without raising annual price. In order to reckon such situations we have analysed changes in monthly prices.

It was hypothesised that both intra as well as inter-year prices are affected by factors on demand side, supply side and by policy interventions. Following model was used to identify factors affecting intra-year and inter-year price variation.

The Model

Changes in monthly and yearly prices of rice and wheat were estimated based on their wholesale price index for the whole country with base 1981-82 = 100. Monthly change in price was estimated by fitting trend to monthly wholesale price index. This was done after identifying the month in which prices were the lowest and the one in which prices were the highest. In the case of rice, prices turned out to be lowest in the month of November and they were highest in the month of September during the study period. Similarly, in the case of wheat, April turned out to be the lowest and February turned out to be the month with the highest price. Monthly growth rate in price was estimated as under:

$$\text{Ln MWPI} = b_0 + b_1 T + u_t$$

Where Ln is natural log, MWPI refers to monthly wholesale price index of the concerned commodity, T refers to the time designating months viz. November to September for rice and April to February for wheat. Monthly trend growth rate in price was given by b_1.

Monthly growth rate in price was multiplied by 12 to get annual rate of inflation in monthly price. These rates of inflation in monthly or intra-year price were used as dependent variables to study the factors affecting growth in intra year prices.

Year to year change in price was calculated as under:

Rice:

[Average of WPI for January to December in year (t) - WPI for year (t-1)]/ [WPI in year (t-1)] x 100

Wheat:

[Average of WPI for April to March in year (t) - WPI for year (t-1)] /[WPI in year (t-1)]x 100

Following variables were tried to explain inter and intra year inflation in prices of rice and wheat.

1. Change in output of the concerned commodity over the previous year (%)
2. Output of coarse cereals - million tonne
3. Growth in output of cereals excluding rice / wheat over previous year %
4. Growth in procurement price over previous year of the concerned commodity %
5. Growth in the PDS issue price of the concerned commodity over previous year %
6. Growth in money supply %
7. Per capita real income in the country at 1980-81 prices
8. Export of concerned commodity – million tonne
9. Import of the concerned commodity – million tonne
10. Buffer stock of the concerned commodity at different times – million tonne
11. Per capita output of the concerned commodity - Kg/year
12. Per capita quantity of the concerned commodity supplied through PDS-Kg/year
13. Net trade of the concerned commodity – million tonne.

Multiple linear regression model was used to estimate the impact of different variables on intra and inter-year price variations. As mentioned above, all the variables having trend in them were expressed in growth rates while variables not exhibiting definite trend were used as such. This way, following forms of regression equations were estimated:

$$\frac{\Delta WPI}{WPI} = b_0 + b_1 \frac{\Delta PRCP}{PRCP} + b_2 \frac{\Delta ISSPR}{ISSPR} + b_3 \frac{\Delta OUT}{ISSPR} +$$

$$b_4\, BUFF + b_5\, EXP + b_6\, IMP + b_7\, PDSPC$$

Where

Δ refers to change

WPI the whole sale price index of the concerned crop

PRCP Procurement price of the concerned crop

ISSPR Issue price of the concerned crop

OUT Production of the concerned crop

BUFF Buffer stock of the concerned crop

EXP Export of the concerned crop

IMP Import of the concerned crop

PDSPC Per capita of the PDS supplied of the concerned crop

Intra Year Price Variation in Wheat

A large number of combinations of different variables were tried to arrive at variables relevant to explain the rate of inflation in monthly and annual prices. In the first run, annual change in wheat output, buffer stock of wheat as on 1st July, export of wheat during the year, change in procurement price over the previous year, change in issue price of wheat in the year and per capita offtake of wheat for PDS were used as explanatory variables. These variables explained 74 per cent variation in monthly price inflation of wheat. In this run all variables except procurement price and offtake for PDS were significant at 5 per cent or lower levels.

In the second round, both the variables, which were non-significant in the first run, were dropped from the equation. In this round R^2 decreased slightly but the adjusted R^2 improved from 0.636 to 0.664. All the four explanatory variables turned out to be significant at 1 percent level. As expected, output growth and buffer stock of wheat caused negative impact on growth rate in monthly price of wheat whereas export and issue price of wheat caused

increase in the monthly price. This equation shows that 1 per cent increase in wheat output reduces monthly rate of inflation by 0.08 % measured on per month basis or by 0.96 % measured on annual basis. On the other hand, 1 per cent increase in issue price of wheat resulted in 1.02% growth in monthly index of wheat prices during a full year.

In the third run, variable representing issue price was replaced by the variable representing annual growth in procurement price. In this run the R^2 declined somewhat, however, the fit was significant at 1% level of significance as revealed by F-test. Again, all the four variables showed statistically significant impact on open market prices. Impact of raise in procurement price was positive and statistically significant at 5 per cent level.

A comparison of results of second and third set reveals that issue price of wheat caused stronger impact on open market prices as compared to the impact of procurement price. For instance, in a 12-month period 1 per cent increase in issue price raised inflation in market price by 1.05 percent point while similar increase in procurement price raised rate of inflation of monthly price by 0.69 percent point. It is also brought out by the analysis that whereas PDS price is significant determinant of variation in monthly wheat price, PDS quantity did not show significant impact.

Inter Year Price Variation in Wheat

Same explanatory variables were used to estimate their impact on change in annual price index of wheat. When all the 5 variables namely percent change in output, issue price and procurement price and level of buffer stock of wheat and its export were used as explanatory variables, R^2 was0.73 and the equation turned to be highly significant. Except procurement price, all variables showed significant impact on annual rate of change in whole sale price index of wheat. Based on this set of regression equation, 1 per cent increase on PDS price of wheat raises open market price of wheat by 0.43 percent point. Output growth has a strong negative impact on year to year price change. One per cent increase in wheat output reduces wheat price by 0.6 per cent. Among other factors, increase in buffer stock of wheat by 1 million tonnes depresses the inflation by 0.78 percent points. This is the main factor for very low rate of inflation in wheat price in the last two years. Export of same magnitude was found to raise rate of inflation by 5.83 points. This is consistent with another study which show that export of wheat exceeding one million tonne destabilise domestic wheat market in India.

In the second set, variable representing issue price was dropped from the model to find out whether impact of procurement price growth could not be captured due to positive correlation between issue price and procurement price. Exclusion of issue price resulted in turning the coefficient of procurement price significant at 2.5 per cent level. Though dropping of issue price reduced the value of adjusted R^2 from 0.650 to 0.50, F-value was significant at 0.3 % level indicating that the fit was good. Durbin-Watson statistics was also not significant, ruling out first order autocorrelation. This equation showed that 1 percent increase in procurement price accelerated the rate of inflation in open market price of wheat by 0.38 percent.

In the third run, instead of procurement price, issue price was used as one of the explanatory variables. In this case, adjusted R^2 turned out to be 0.647, which is almost same as in the first run. The results of different set of equation show that issue price is a much stronger determinant of market price as compared to the impact of procurement price on open market price.

Intra Year Price Variation in Rice

After a large number of trials with different variables, two sets of equations were retained to explain intra year variation in prices of rice. In the first set, five variables were used as predictors. The equation explained 59 percent variation in monthly prices of rice. All the variables had expected sign but the impact of issue price of PDS rice and output growth did not exhibit statistically significant influence upto 10 percent level on monthly price variation.

In the second round, growth rate in issue price of PDS rice was dropped and other variables including quantity of PDS rice were retained. This did not make any difference to the explanatory power of the equation. In this set, significance level of all the retained variables improved, however, impact of output growth on the monthly open market prices turned significant only at 17 percent level. This equation seems to explain best the inflation in monthly price of rice during the period 1977-78 to 1998-99. Based on the results of this equation, one percent increase/decrease in output of rice reduces/raises rate of inflation in monthly prices by 0.13 point. Impact of changes in procurement price was quite strong - monthly prices rise by almost same level (1.06 point) as the change in procurement price. An increase of 1 kg rice per capita for PDS reduces the rate of inflation in monthly price by 0.187 points. Export of rice results in increase in open market prices and the impact was significant at 2.7 percent. Export of rice to the tune of 1 million tonne raises intra year price inflation by 3.16 percentage points.

The reason for output growth not showing significant impact on intra year price rise seems to be lower level of seasonality in its production. Unlike wheat, growing season and harvest of rice are not confined to one season. Though in north India rice is grown during mid June to October, in southern and eastern parts of the country its cultivation is not confined to these months alone. This is also evident from the monthly market arrival pattern of rice. For instance, figures for the year 1995-96 shows that 65 percent market arrival of wheat is concentrated in the quarter April to June as compared to 56 percent in the case of rice in the months of September to December.

Variables like import and buffer stock of rice were not found relevant to explain monthly variation in rice prices. Spread of rice cultivation throughout the year could be the reason for buffer stock of rice not showing significant impact on the price variations.

Abnormal Price Rise in the Wake of Normal Production

There have been instances when prices of rice and wheat used to rise very high despite normal or even above normal domestic production. These have often caused concern as it generally used to be believed that such instances reflect poor management of food economy. Blame has also been put on private trade for manipulating market to take advantage of high off season rise in prices which has not been justified by factors on production side. There is need to identify factors which gave rise to this kind of situations.

Intra and inter year price variations are caused by several factors and their interacting influences. However, it is instructive to identify specific situations and underlying factors when prices go abnormally high despite normal production.

In the first step, years showing abnormal or excessive increase in wheat price need to be identified. This requires a norm that can be considered as normal. One way to do this is to take average and standard deviation of the concerned observations and to classify those observations as abnormal whose deviation from mean exceed standard deviation. This way, during last 22 years covering period 1977-78 to 1998-99 intra year price rise in the case of wheat was abnormal during 1988-89, 1990-91, 1991-92 and 1996-97. There are also instances to use some subjective norm to identify abnormal years. This is generally done by choosing some arbitrary norms. For example, if we consider monthly growth rate in prices above 2% as abnormal, it would capture situation which deviates more than 35% from the average. Based on this, 2 more years namely 1980-81 and 1982-83 also turn out to be the abnormal years during which intra year price spread exceeded 24%.

Out of these six years identified to experience abnormal price rise, production was below trend during 1980-81, 1990-91 and 1996-97, which substantially contributed to high rate of intra year price inflation. In the remaining 3 years there were factors other than production, which are responsible for high rate of price increase.

In 1981-82 procurement price for the first time was raised by 11.1% whereas average rate of growth in procurement price during the chosen period was 7.33%. Another factor, which seems to have contributed to high price rise is low level of buffer stock. Other two factors responsible for high price rise in this year are low capita supply of PDS wheat and above average increase in its issue price.

During 1988-89, buffer stock of wheat in July 1988 had touched level of 7.6 million tonne, which was lowest during the previous decade. This is the main factor for abnormal price rise during 1988-89. Beside, stagnant production and high export in previous year have also been favourable for price rise of high magnitude.

In the case of rice, average rate of growth in monthly prices during the last 22 years was about 1 percent and its standard deviation was 0.621 percent. Growth in monthly prices exceeded this average plus standard deviation in only 3 years viz. 1979-80, 1981-82 and 1991-92. This does not provide adequate number of observations to identify factors other than output responsible for sharp price rise. In order to have more number of observations to study this phenomenon, price rise exceeding 1.25 percent monthly growth or 15 percent annual growth was considered as abnormal price increase year. This way 7 years were identified as abnormal years. This includes 1977-78, 1979-80, 1980-81, 1981-82, 1983-84, 1989-90 and 1990-91. In addition to these years 1988-89 and 1992-93 were also considered as abnormal years due to abnormal inter year price rise. This way 9 out of 22 years showed abnormal price rise. Out of these, rice output was below normal during 1979-80 and 1992-93, while per capita PDS supply was lowest during 1979-80 and output of coarse cereals was lowest in 1992-93 in the entire period of study – these explain high rise in rice prices. Increase in procurement price turned out to be the major factor for abnormal price rise in most of the years. During 1980-81, 1983-84 and 1992-93 low output of coarse cereals also contributed towards sharp rise in rice price.

Temporal and Spatial Price Band for Private Sector Participation

India's food grain sector has undergone tremendous changes since the launch of green revolution technology during late 1960s. The most profound impact has been witnessed in the case of rice and wheat whose production has shown spectacular increase and enabled the country to move from chronic food shortages to an era of food surplus. Besides technology and increased use of inputs, other factors like government intervention and support in terms of guaranteed minimum prices for output, and large scale procurements from producers at remunerative prices have played key role in growth of food grain output particularly of rice and wheat. These government interventions have been affected through parastatals like Food Corporation of India (FCI) which procures rice and wheat from surplus regions, maintains their stock, supplies rice and wheat for public distribution system in various states and undertakes open market sales. This intervention has been quite effective in achieving goal of food security for the country, but its sustenance has been putting large burden on the state exchequer in the form of food subsidy which have been growing year after year.

Government is increasingly finding it difficult to bear the burden of mounting food subsidy; and questions are being raised whether the benefits of food subsidy are reaching the poorer sections of the society. It is being suggested that food subsidies can be decreased considerably by encouraging participation of private trade in food grains. These suggestions follow from the fact that the country is in a comfortable position on production front and a sound network of roads and markets have been developed for efficient functioning of private trade. On the other hand, there are fears of the past that private trade would not pay remunerative prices to producers and would manipulate market to exploit consumers. It is also feared that withdrawal of government intervention in grain markets would jeopardise food security. This underscores the need to evolve new kind of mechanism for government intervention in food grains markets which allows and encourages active participation of private trade but keeps a check on exploitative tendencies of private trade. This can be done by creating an environment that provide reasonable incentive to private sector to operate in food grain market. This implies that as long as private trade charges normal margin for delivering the commodities across space and time, government would not intervene in the market. Implementing such mechanism would require developing norms which can be used to find out whether private trade is charging reasonable prices or is indulging in exploitation of consumers and producers. The study prepares estimates of price spread between farm harvest prices, wholesale prices and retail prices of rice and wheat. By reckoning all relevant costs like statutory charges, labour charges, processing charges, transport cost, storage cost and cost of working capital, and reasonable margins for different functionaries, the study develops estimates of price band between farm harvest prices of wheat and paddy in surplus states and wholesale and retail price in subsequent months in all major states of the country. The acceptable price band includes reasonable margin for wholesalers, rice millers and retailers, and, marketing costs like (1) statutory charges, (2) labour charges, (3) material cost, (4) local transport and freight to different states (5) milling charges for rice (6) storage charges and (7) interest on working capital.

Methodology

The price band has been estimated for rice and wheat which are most important food grains for the country from production as well as food security point of view. The band corresponds to grain trade that takes place between surplus and deficit states. A state has been termed as surplus or deficit depending upon the difference in per capita net production and consumption during the recent triennium. Per capita net production of rice and wheat in each state has been taken after deducting 12.5 and 10 per cent of output of wheat and rice respectively as allowance for feed, seed and wastage. Estimates of per capita consumption have been taken from 50th round of N.S.S pertaining to the year 1993-94.

Price band has been computed in two steps by taking farm harvest price as the lower threshold. In the first step price band has been computed between farm harvest price (price paid to producers by buyers in primary market) and wholesale price. In the second step the band has been extended to retail level. This implicitly assumes following type of marketing channels for rice and wheat in deficit and surplus states:

Wheat (surplus states)

Producer —— wholesalers in primary market —— retailer

Wheat (deficit states)

Producer —— wholesaler in primary market —— wholesaler in secondary market —— retailer

Paddy/Rice (surplus States)

Producer —— Wholesaler cum rice millers in primary market —— retailer

Producer —— rice miller —— wholesaler in primary market ——retailer

Paddy/rice (deficit states)

Producer —— rice miller —— wholesaler in primary market—— wholesaler in secondary market —— retailer

The trade flows take place like this. In the case of wheat wholesaler buys wheat from farmers in the primary market in surplus state and pays (i) statutory charges like mandi fee and commission (ii) cost of labour and material. When these cost are added to farm harvest price it becomes wholesale price in the surplus state for the harvest season. Taking the produce to deficit states during harvest season would involve transport cost, margin of wholesaler in the primary market and freight charges. When these costs are added to wholesale price in surplus state it gives the price paid by wholesaler in the deficit states. It is assumed that during harvest season transactions are instant and that there is no storage and no capital is blocked in the business.

Wholesale price in the subsequent months is derived by adding storage cost, interest on working capital and wholesaler's margin. The storage cost and interest on working capital are charged for half of the duration of harvest season i.e. 45 days and for the duration of period after harvest season. For the deficit states transport charges and forwarding charges are also added. Forwarding charges include labour charges for loading the produce in trucks to despatch to other destinations. In the case of rice, farm harvest price of paddy in surplus region is used as the lower brand.

The second step involves estimation of price spread between wholesale and retail price. It was assumed that difference or ratio of long run average of retail and wholesale prices of comparable grade in each state would represent normal price spread which includes wholesalers margin, retailer's margin and costs included in movement of produce from wholesale to retail level in the same state. Based on this, average of percent difference between wholesale and retail price of rice and wheat for same grade for the last 12 years (1986-98) was used as price spread between wholesale and retail. Due to lack of consistent data series, price spread between wholesale and retail in some states could not be estimated. In such cases estimates for other grain i.e. wheat for rice or rice for wheat were used.

Each state was identified as surplus or deficit in rice and wheat based on the difference between per capita net production and consumption. Punjab, Haryana, Uttar Pradesh turned out to be surplus in both rice and wheat. Among other states, Rajastan and Himachal Pradesh were surplus in wheat while West Bengal was surplus in rice.

Out of the surplus states, Punjab was chosen to represent surplus region as highest surplus is available in this state. Within Punjab, Khanna market was selected to represent farm harvest prices and wholesale prices in the surplus state. Detailed information on marketing costs and margin, processing cost and transportation cost was collected from various market functionaries in Khanna market. These functionaries include traders, commission agents, wholesaler, miller, transport companies, and officials of market committee.

Price Band Estimates for Wheat

Farm harvest price of wheat during April-June, 2000 was taken as lower band of price. This price was reported to be Rs.582 in Khanna market in Punjab. After purchase of wheat the seller pays purchase tax (4%), surcharge (1%), market fee and rural development fund (4%) and commission (2.5%). These are termed as statutory charges and they sum up to 11.5 per cent of the price paid by wholesalers to producers in assembly market. Wheat is then filled in bags of about 95 Kg weight, which are then stitched and weighed and then transported either to godowns for storage and sale in subsequent months or despatched directly to secondary markets mainly in deficit states. Wholesaler bears labour charges for filling, stitching, weighing and loading and unloading and charges for transporting produce to godown. Wholesaler also pays for packaging material (jute bag) which is generally used twice. When all these costs are added then reference wholesale price in surplus state comes to be Rs.666.42 per quintal.

Output stored for sale in subsequent months is charged Rs. 1 per bag per month as storage charges and interest on working capital. Wholesalers in Khanna market reported that they get finance from two sources viz. private and institutions. They pay 14 per cent annual interest on institutional advances and 18 per cent on private advances; we have used average of interest rate of two sources i.e. 16%. Subsequently when produce is released for sale in lean period, wholesaler incurs cost of Rs. 1.00 per bag as loading or forwarding charges. According to the wholesalers in Khanna mandi they require a margin of Rs. 20 per bag as profit or incentive to remain in the business and to meet their establishment costs. This level of margin is close to the actual margin realised by wholesaler in wheat trade which is reported to be Rs. 25.10 per quintal according to a World Bank study.

Transportation charges particularly by road (truck) keep changing frequently. Transport cost per quintal of foodgrain between surplus state (Khanna market) and other states as applicable during April - September 2000 is given in Table 1.

Table 1 :Transport cost per quintal produce by truck to various states from surplus state (Punjab)

State	*Market*	*Freight*
Andhra Pradesh	Hyderabad	261.00
Assam	Gwuahati	297.50
Bihar	Muzzafarpur	216.50
Gujarat	Anand	122.00
Haryana	Karnal	41.50
Himachal Pradesh	Simla	75.00
Jammu & Kashmir	Srinagar	186.00
Karnataka	Hubli	247.00
Kerala	Ernakulam	286.50
Madhya Pradesh	Indore	127.50
Maharashtra	Mumbai	191.50
Orissa	Cuttack	161.50
Punjab	Khanna	
Rajasthan	Kota	67.00
Tamil Nadu	Chennai	275.00
Uttar Pradesh	Kanpur	111.00
West Bengal	Calcutta	247.00
Goa	Panjim	275.00
Delhi	Delhi	50.00

Marketing cost, mandi labour and material charges comes out to be Rs.84.43 per quintal of produce and they do not vary over months. Storage and interest charges for half of the harvest season and for half of the following month are included in the month following harvest season. This way first month following harvest season includes interest and storage charges for two months time. For the next 8 months monthly charges are added to the July charges. This way marketing costs and margin are estimated to increase from Rs. 105.42 per quintal of wheat during April- June to Rs. 204.39 in March next year. Adding all these costs and transport cost to farm harvest price gives upper price band for wholesale price.

Corresponding to farm harvest price of Rs. 582, wholesale price of Rs. 666 in harvest period to Rs. 786 in March next year provides reasonable incentive to private sector to undertake marketing of wheat in surplus states. The range is higher for deficit states. For instance, wholesale price between Rs. 948 - Rs.1052 in different months is reasonable for state of Andhra

Pradesh. Wholesale price exceeding this range would indicate that private trade is exploiting market situation to earn excessive profit. Upper price band for wholesale price of wheat in Delhi market comes to Rs 737 for April to June, and it extends to Rs. 836 by March next year.

The second stage of price band between wholesale price in a state and the retail price has been estimated by multiplying wholesale price by average of the ratio of retail to wholesale price for the last twelve years beginning 1986. This long run average assumes that the abnormal profit/losses cancel out in the long run and that long run average presents normal margin. These estimates show that retail price of wheat in the country were higher by 15.46 per cent of the wholesale price. Minimum price spread was observed in the case of West Bengal where retail price of wheat exceeded wholesale price only by 5.27 per cent and highest price spread was observed in the case of Maharashtra where the retail price of wheat was 24 per cent higher than the wholesale price in the state. Major reason for variation in price spread between wholesale and retail level seems to be due to differences in tax structure. For some of the states consistent price data was not available to estimate spread between prices. In such cases price spread of other food grains i.e. wheat for rice or rice for wheat was used.

Table 2: Price spread between wholesale and retail prices of foodgrains 1986-98; Unit: percent

State	*Rice*	*Wheat*
Andhra Pradesh	14.40	—
Assam	17.05	—
Bihar	10.55	15.77
Gujarat	—	23.04
Karnataka	7.11	—
Kerala	10.76	—
Madhya Pradesh	8.88	10.52
Maharashtra	12.26	24.09
Orissa	8.37	—
Punjab	—	18.51
Rajasthan	—	13.69
Tamil Nadu	16.03	9.66
Uttar Pradesh	14.36	13.25
West Bengal	7.04	5.27
Delhi	15.12	20.84
Average	11.83	15.46

Retail price band turns out to be lowest for the state by Rajasthan where it ranged between Rs.758 to Rs.894. Highest retail prices were observed in the case of Goa. During 2000-01, private trade earns reasonable margin by selling wheat in retail at Rs. 891 during April to June to Rs. 1011 in March next year in Delhi market.

Private trade would have incentive to undertake marketing of wheat by realising retail price 30 to 105% higher than farm harvest price. In wheat surplus states retail prices that are 30 –36 per cent above farm harvest prices would be profitable for private trade. The margin increases with the distance of deficit states from surplus states and time difference after harvest season. By the month of March next year retail prices in Andhra Pradesh, Assam, Kerala, Maharashtra, Goa and Tamil Nadu must be double or more than farm harvest prices in surplus states to provide incentive to private trade to remain in wheat trade.

Price Band estimates for Rice

One major difference between rice and wheat marketing is that the latter is traded mainly in the grain form purchased from farmers while former involves processing/milling of paddy to convert it into rice. Farm harvest price of paddy (fine grade) has been used as lower threshold for computing corresponding wholesale and retail prices of rice. In the trade flow it is assumed that farmers sell paddy which is purchased by rice miller who in most of the cases also happens to be rice wholesaler. Like wheat, miller/ wholesaler pays 11.5 per cent of the purchase price of paddy as statutory charges. Other charges paid by the wholesaler cum miller in the market are not proportional to value of produce but are paid on per bag basis which consists of about 60 kg paddy. For one quintal of paddy these charges are estimated at Rs. 3.85 for bag filling and stitching, Rs. 1.54 for weighing and Rs. 5.77 towards loading and transport to rice mill. Paddy wholesaler also pays for cost of gunny bag. We were told that one gunny bag which costs Rs. 20 can be used thrice for paddy after which its junk value would be Rs. 2. Thus cost of packaging material comes to be Rs. 6 per bag or Rs. 10 per quintal of paddy. All these costs sum up to Rs. 80.96 per quintal of paddy.

After its purchase, paddy is processed into rice. Accordingly to the information furnished by rice millers in Khanna market, cost of processing one quintal paddy is about Rs. 26. Rice millers recover a large part of this cost from sale of rice husk and rice bran. Based on our sample information rice bran and husk derived from milling 100 kg paddy gives a return of about Rs. 19.60 depending on the price realised for bran and husk. Thus, net cost of processing 1 quintal paddy comes to Rs. 6.40 (Rs. 26 less Rs. 19.60). This cost was further used to arrive at estimate of rice miller's margin. Rice millers process paddy to rice for Food Corporation of India by charging Rs. 13 per quintal of paddy under custom milling arrangement. Based on this, net margin to miller turns out to be Rs. 6.60 per quintal of paddy.

According to information furnished by rice millers to us genuine cost of processing one quintal paddy comes to be Rs. 26. Compared to this a study done for World Bank by Piggot and Sidhu finds this cost for year 1998-99 to be Rs. 26.09 when land rent is not considered.

After milling, rice is stacked/stored which involves cost of labour and gunny bag. There is also cost of capital which is estimated at the rate of 16 percent for half of the duration of harvest season. As one quintal of rice is extracted from 1.5 quintals of paddy, all these costs and cost of one quintal paddy are multiplied by a factor of 1.5 to get cost of one quintal rice to wholesaler. This way wholesaler's price for rice turns out to be Rs. 952.97. When this rice is further sold by wholesalers to retailers in surplus states or to wholesaler in deficit states it would involve wholesaler's margin of Rs. 20.27.

Policy Options and Strategy for Foodgrain Management

Foodgrain management operations of government have recently come under increasing criticism due to various problems associated with such operations and in the context of prevailing environment for policy reforms. These operations include government intervention in maintaining buffer stock for food security and price stabilisation, procurement and distribution of foodgrains and farm price policy. It is contended that all these interventions were justified in the time of food shortages and scarcity and when marketing infrastructure and institutions were not sufficiently developed and competitive to protect interest of producers and consumers. Easing of strain on domestic food availability, accumulation of massive stock of cereals, emergence of shortages in edible oils and pulses and new options for price stabilisation are cited as strong reasons for re-looking into the entire policy of food management and government intervention in foodgrain markets.

Inter Year Price Stability

Cereals being the basic and staple food were considered as the most important component of food security. Their demand and supply are both inelastic implying that small fluctuation in output is likely to cause large increases/decreases in their prices which in turn would result in sharp fluctuations in farm income and expose consumers to high volatility in food prices. Thus, in order to bring stability in prices and farm income it was considered imperative to maintain buffer stock of grains which involved purchases from the market during the good harvest years and releasing stock during lean years or when production is below envisaged trend.

Purchase from market and release of stock in this manner ensures that supply of produce in the market would move on a smooth trend. If growth in demand keeps pace with the envisaged trend in supply then purchase/ release of stock based on deviations from trend level of production would ensure perfect balance between demand and supply assuming closed market from external trade. This should ensure that there is no disturbance to price stability as for as genuine factors on demand and supply side are concerned.

During 1970s highest positive deviation from trend level of production (henceforth also referred to as normal production) was 12.81 million tonne and highest negative deviation was 15.31 million tonne. Thus, during the decade of 1970s highest procurement of 12 81 million tonne and buffer stock of 15.31 tonne was required to take care of fluctuations in supply. Procurement of cereals during this decade varied between 6.25 and 14.46 million tonne.

During 1980s actual production exceeded normal level by 8.52 million tonne during 1983-84 which was highest for the decade while highest negative deviation was experienced during 1987-88 when cereal output declined from normal production by 16 million tonne. Procurement during this decade varied between 12.31 and 23.13 million tonne.

During the decade of 1990s fluctuations in cereal output declined considerably as deviations from normal production remained below 7.06 million tonne on either side. However, procurement by official agencies during 1990s rose sharply and ranged from foodgrain 16.23 to 31.53 million tonne. Like the increase in government procurement of foodgrain, level of buffer stock of cereals also moved up over time. This shows that reduction in instability in cereal production in the country was not followed by decline in quantity of output purchased by government and buffer stock. To be more critical, quantity of foodgrains purchased by government and level of buffer stock have not been dictated or guided by fluctuations in output. If maintaining inter year price stability was the important consideration for government intervention in grain markets then reduced fluctuations in output, experienced in the country, should have resulted in decrease in level of procurement and buffer stock rather than following increase over time.

Assuming that private trade would not play major role in maintaining inter year price stability, government can base its decision to buy above normal output on advance estimates of production available around harvest time. This would be given by deviation of estimated output from stipulated trend, which can be estimated quite reliably for medium term. Similarly, level of buffer stock for such purpose can be worked out from negative deviations of actual output from stipulated trend. During the last decade highest decline from trend level was observed to be 7.63 million tonne. Also, after 1987-88 India did not face two consecutive bad years of cereal production. It is thus safe to assume that buffer stock of around 7.63 million tonne is sufficient to make up for the deficiency in cereal output from normal level.

There would be instances of severe drought like one experienced during 1979-80 and 1987-88 when decline in output could go quite high. Should buffer stock be maintained to take care of such serious shortfall which occurs once in a while in about 15 years period? This kind of events are highly unpredictable about their severity as well as occurrence. It does not seem to be prudent to link level of buffer stocks to magnitude of output decline in such years as it includes maintaining very huge level of buffer stock for a very long period. Normal buffer stock would take care of about half of the output shortfall in the time of severe droughts. Some of the output deficiencies can be met from international market and some can be compensated for by reduced consumption. As far as procurement from domestic markets for price stabilisation is concerned, it is suggested that the Centre should carry out open market purchases and sales from the market.

Supplies for PDS

Foodgrains are also procured by government to service Public Distribution System and other welfare schemes of the Government. Strong debate is going on about the desirability of continuing present system of public distribution and suggestions have been made to replace it with more efficient "food stamp" system (Ramaswami 2002 and Gulati et al 2000). Here, we are not going into the merits and demerits of PDS and feel that for the time being some sort of PDS would be needed as a safety net for vulnerable sections of the society.

Total requirement for PDS is estimated to be 21 million tonne per annum (Economic Survey 2000-01, p. 50). Procurement for PDS would be a continuous exercise requiring purchase of rice and wheat each year. This is in sharp contrast to the procurement to meet inter year stability in supply which should arise only when anticipated output is above the trend or normal level of output. Thus, total procurement of grain in a given year should be around the sum of quantity needed for PDS and excess of estimated output from normal production. Similarly, maximum level of buffer stock should be 18.13 million tonne which is sum of stock needed for PDS for six months (21 divided by 2) and highest shortfall observed in grain production during the recent 10 years (which is found to be 7.63 million tonne).

Implementing Guaranteed Prices

Besides maintaining inter year price stability and PDS supply, third important goal of government intervention in grain markets has been to ensure guaranteed prices to producers as announced by government before harvest season. This is ensured by buying all grains delivered at the support price. If market price is above support price then there is no need for procurement by government to ensure guaranteed price.

Here it is important to understand implications of guaranteed purchases to ensure minimum support price for the produce. If support price is consistent with demand and supply for the given commodity then competitive market should ensure that market price does not go below the level of support price. To elaborate it further, if difference between support price and subsequent open market price leaves adequate margin for private trade then, under competitive markets, private trade would be offering price to producers higher than the support price. In such a situation there would be no need for government procurement to ensure minimum price to producers. However, if market is not competitive or is unfavourable for private trade then there could be need for government to ensure minimum support through purchases from the market. Thus, when market price in the lean months, as determined by demand and supply, leaves enough margin above support price then support price can be ensured through competitive market and trade without government procurement.

In case the market price determined by supply and demand does not leave adequate margin over support price, private trade would not buy the produce at support price howsoever competitive the market may be. This can happen when growth in demand does not keep pace with the supply and, when support price is fixed based on supply side factors only. This is exactly what is being experienced in the case of rice and wheat in the country.

During 1990s procurement prices and wholesale prices of rice and wheat increased at a much faster rates than the growth rate in general price index. Similarly, the rate of inflation in rice and wheat prices during 1990s was significantly higher compared to the 10 years period preceding 1990-91. Rural retail prices and PDS prices, which actually matter for consumption, also increased at a much faster rate during 1990s compared to 1980s. Thus both open market purchases and PDS demand have been affected by the high growth in prices causing a decline in per person purchase from PDS and open market.

Beside prices there are indications of cereal demand getting reduced to some extent due to changes in consumption pattern. This is evident from the fact that despite the lower growth in retail prices of cereals compared to prices of other food, implying decline in real terms, per capita cereal demand has followed declining trend. This way, both price factor as well as change in consumers preference to shift from cereals to non cereal products are responsible for decline in per capita cereal demand in the recent years.

To the extent deficiency of cereal demand is resulting from long run changes in consumption pattern, prices would not be very potent instrument for boosting demand and for restoring balance between demand and supply. The solution lies in adjusting supply to demand. One way to do this is to bring some reduction in area under cereals in inefficient producing regions.

Price based solution to restore balance between cereal supply and demand in the present situation requires market prices to come down to raise the level of demand and reduce the level of supply till they are in equilibrium. However, it has very serious implications which may not be desirable from long run food security point of view. Taking year 2000-01 as an example, purely price based solution implies that to maintain balance between supply and demand, cereal prices should have been lower by 27 percent. It is worth noting that year 2001 was not agriculturally good year and cereal output was about 7 million tonne below normal level based on trend. Despite that and despite net export of 2.9 million tonne, cereal stock increased by 12.5 million tonne in the year 2001. If demand for cereal remains deficit and cereal output increase to their normal (trend) level then required decrease in price to maintain balance between supply and demand works out to be enormous. This kind of purely price based solutions would result in steep decline in crop income and would cause adverse impact on agriculture sector which is already under threat due to trade liberalisation and, is not able to cope up with low level of international prices.

The best way to ensure remunerative prices to foodgrain producers and reducing unwanted surplus in future is to take measures that result in some shift in resources from cereals to non cereal enterprises and encourage growth of cereal output in efficient producing regions. This strategy should not be based only on reducing profitability of grain production by lowering their prices but it should involve providing and developing alternatives which are more remunerative than cereals.

There is also a need to examine that if the level of guaranteed prices based on a particular criterion is encouraging production that society can't afford to buy then what other criterion should be used to fix guaranteed price. This change is also important as experience shows that it is not feasible to enforce guaranteed prices everywhere in the country. Enforcement of such prices requires mechanism for procurement, which now exists only in a few states. In rest of the markets agricultural produce is purchased by private trade. If private trade has a feeling that guaranteed prices are too high to be supported by demand side factors, it would not buy produce at that price. This is what has been experienced in the case of rice and wheat in the last 2-3 seasons. Actual prices received by farmers are reported to have remained lower than guaranteed prices in several markets other than those where government procured the produce.

Regarding implementation of guaranteed prices, some important lessons have been learnt in the recent years. Mere announcement of higher support prices for commodities, which are not effectively backed up by procurement arrangement, does not serve the purpose of raising level of prices received by producers. Therefore, attempts by CACP to raise support prices of crops like edible oils and pulses, in which India is deficit, relative to support prices for rice and wheat which are in excess supply, cannot be expected to result in shift of resources from rice and wheat to the deficit crops.

Dealing with Grain Stocks

During the last five years level of cereal stock has accumulated to such a high level that it has put the government in a very embarrassing situation besides putting huge financial burden involved in carrying these stocks. Several reasons have been given for accumulation of grain stock much above the genuinely required level. The foremost reason is said to be hefty hike in procurement prices and consequent increase in open market wholesale prices of rice and wheat which then resulted in reduced demand. In fact several factors combined together to build up so called "mountains of grains" in the country.

First, during the decade of 1990s procurement prices of rice and wheat were given a comfortable increase which was quite higher than the increase in general prices. This was followed by steep rise in procurement by government agencies in surplus regions. The increase in procurement prices necessitated increase in issue prices and, because substantial part of marketed surplus was with the government, its policy on quantity and price of stock released had a determining influence on open market wholesale prices of rice and wheat. The influence on wholesale prices was such that they increased by even higher rate then procurement prices. As demand side factors did not support this increase in price, growth in retail prices started trailing behind the growth rate in, largely government influenced, wholesale prices. This caused adverse impact on the margin of private trade which slowly started withdrawing from the market. Accumulation of cereal stock with government agencies created a feeling that release of excess stock can anytime depress open market prices. These two factors led to withdrawal of private trade from grain markets in surplus states causing increase in procurement by government agencies. Another reason for reduced role of private trade in direct purchases from producers is the release of stock for open market sale and export at a much lower price than what would be the unit cost of rice/wheat to private trade from direct purchases from producers. This created perverse incentive to private trade - not to participate in primary market and buy from government rather than buying produce from producers.

Second, increase in PDS and open market prices and diversification in food consumption habits led to demand deficiency resulting in diversion of production to inventory.

Third, steep decline in international prices of cereals since 1997 has caused adverse impact on export of cereals and excess stock could not be smoothly sold in international market.

Unless excessive stocks are liquidated, grain markets in India would continue to remain in trouble. There is no single and easy way to do this and several options need to be tried. Some of such options are already suggested in the Economic Survey 2001-02 (p.127) which need to be pursued seriously. These are (i) increasing BPL quota (ii) expanding employment generation programmes (iii) universal PDS at BPL prices in calamity affected areas. In addition to these, some other options also need to be tried. Very old stock that has deteriorated in quality should be disposed off as a feed in international and domestic market at whatever price it can fetch. If still excessive stock remains, some of it should be given as food aid to needy countries.

Trade as an Alternative to Buffer Stock

Buffer stocks have been used by the government as an important instrument for the purpose of price stabilisation. However, this involved heavy cost in terms of procurements, handling, carrying, storage etc. which is becoming fiscally unsustainable. As an alternative it has been suggested that government should use the instrument of variable levies on external trade to stabilise domestic prices. Under this, it is suggested that when international prices are low, tariff on import should be fixed high to provide price support to domestic producers and when international prices go high, tariff should be reduced. Similarly, variable levies on export can be worked out for net exporting countries. These authors found the trade option to be superior compared to buffer stock in stabilising prices under liberalised trade regime.

On the other hand, it is contended that due to its large size of demand and supply, India can't rely on international market to meet its demand for staple food. It is often said that meeting grain deficits during lean years and selling surplus during years of bumper harvest in international market for domestic stability would destabilise international prices. Thus, international prices would turn out to be much higher during import and they would go very low during export. Thus trade option is believed to be appropriate for counties or products with small demand/supply.

In order to examine the implications of trade options for domestic price stabilisation in India, we have estimated the changes in international prices when India sells/buys its surplus/deficit grain in international market. The impact of India's export and import on international price of wheat assumes that price elasticity of wheat prices with respect to changes in quantum of world trade is unity. Similarly, the impact on rice prices is estimated based on the assumption that changes of one million tonne export of rice by India results in 4.7 percent change in international price of rice. It was further assumed that inter year price stability requires purchase by government exceeding trend output and sale from buffer stock to the tune of deficiency of output from the trend output.

The impact of buying deficit and selling surplus output in international market on international prices can be seen by comparing international prices presented in columns 2 and 6. For instance, for year 1975, wheat output fell short of stipulated trend by 2.77 million tonne. When this quantity is purchased from international market it raised international price of wheat from Rs. 1303 to Rs.1358 per tonne. During the period 1975-2000, wheat production exceeded trend in 10 years and it was below the trend in 16 years. The highest shortfall during this period was recorded in year 1980, when output was short by 3.83 million tonne. Buying this entire quantity from international market raised international price nearly by 4 per cent. On the other hand, highest positive deviation was observed during the year 2000. Disposing off entire surplus of that year in international market reduced international price by 4.76 per cent. This shows that stabilising

India's domestic supply of wheat through sale and purchase from world market did not cause very high impact on world prices The reason for this is that global trade in wheat is quite large and deviation in India's output constitutes less than 5 percent of the global trade. This shows that the fear of very high fluctuation in international price due to India's dependence on International market for price stability (for wheat) is misplaced.

The next question is whether trade option could be superior to the policy of domestic stabilisation through buffer stock. This comparison involves lot of complications and several assumptions. To accomplish this, we have estimated another set of trade prices named as import parity prices

(IPP) and export parity price (EPP). Import Parity Price for wheat is sum of following:

a) International price for wheat (US HRW2, FOB US gulf ports)

b) Ocean freight rate from US to India

c) 5 per cent value of FOB price as port clearance and marketing charges.

d) Cost of internal transport from port to consuming region as wheat is mainly consumed in the non coastal states.

Import parity price computed this way is comparable with economic cost of FCI for the country. Similarly, export parity price is computed by subtracting following cost from FOB wheat price for US HRW wheat adjusted for India's export: i) Cost of transport from producing state to port. ii) Cost of produce handling, marketing and port charges taken as 5 per cent for domestic wholesale price.Wheat price in Hapur market was used as a reference price for this.

A somewhat crude measure of gain/loss to consumers from government intervention in maintaining domestic price stability is given by the difference between import parity price and economic cost of FCI for importable situation. Similarly, difference between export parity price and domestic wholesale price for exportable situation is considered as the loss/gain to producers from government intervention for price stability. This comparisonfor wheat shows that out of 16 years, when domestic supply was short of trend, cost of meeting the supply deficit from domestic sources (economic cost to FCI) was lower than import parity price in ten years. In 6 years, meeting shortfall in supply from import turns out to be cheaper option for maintaining stability in domestic supply.

If domestic wholesale price is assumed to be the outcome of government policy of price stabilization, then its comparison with the net price that can be earned from sale of produce in international market during the years of above normal production indicates gain/loss to producers from domestic price stabilisation. Wheat production was above normal in 11 out of 26 years. Out of these 11 years, price realised from export turns out to be lower in as many as 10 years. Selling in international market would have fetched better price than that available under government intervention in only three years.

In the case of rice, price of permal rice in Delhi market was used as domestic reference price for export, which is comparable with price of 5 percent broken Thai rice. However, economic cost of FCI for rice refers to common rice which is generally of a poor quality and is comparable with 35 per cent broken Thai rice. Thus, for comparing import parity price (IPP) with economic cost of FCI rice, price of 5 percent broken Thai rice was discounted by 20 per cent which took that price closer to the price of 35 percent broken rice. For importable scenario, ocean freight rate from Bangkok to Indian port and 5 percent cost of marketing, handling and port changes was added to discounted price of 5 per cent Thai rice. But, domestic transport cost was not added to it as coastal states are the major consumers of rice.

World trade in rice is quite thin compared to the volume of wheat trade. Therefore, India's export/import corresponding to deviation in rice output caused large changes in international price. For instance, in the year 1975, when rice output was 3.74 million tonne below the trend, meeting this deficit through import raised international price from Rs. 3051 per tonne to Rs. 3516 showing an increase of 15.2 percent. In the case of rice production, India sometimes faced severe supply shocks reaching the level of more than 9 million tonne. In such cases, India's import to meet the output deficiency raised international price by more than 39 percent. Positive deviations in the case of rice were quite low compared to negative deviations.

A comparison of domestic stabilisation measures and trade shows that during the 12 years since 1975, when output was short of trend, economic case of rice with government agencies was lower than IPP in as many as 10 years. Only in 2 out of 12 years import was more cost effective option than domestic stabilisation.

In the case of second scenario, when actual output exceeded the trend, domestic producers could earn better from export only in five out of 14 years. In the remaining 9 years domestic wholesale price as determined by government intervention turned out to be higher than export parity price.

The above analysis shows that selling and buying wheat in international market to stabilise domestic output does not result in large changes in international prices of wheat due to large volume of world trade in wheat. However, in the case of rice, stabilising domestic supply through trade caused sharp fluctuation in international price of rice. Among the two options viz. domestic stabilisation through buffer stock and stablisation through trade the latter is found to be costlier than domestic stabilisation in most of the years though it also depend upon fluctuation in international price. If the relationship between domestic and international price in future remains same, as observed during the last 26 years, than policy of price stabilisation through buffer stock seems to be better option than trade.

Performance of FCI

Food Corporation of India has remained in the centre stage of government intervention in agricultural marketing due to scale of its operation and due to its role in food security. Though this agency has played significant role in ensuring guaranteed price, and hence in adoption of improved technology in traditional green revolution region, its cost of operation and efficiency have remained subject of criticism and are seen as the main factor for mounting food subsidy in the country. Some studies find that FCI cannot be blamed for high cost of foodgrain handling and distribution and economic cost of its operation turns out to be lower than the private trade if latter pays same statutory charges and serve the same purpose which former has been doing. Similarly, Madhura Swaminathan also find that costs of FCI were justified and that this organisation is vital for food security of the country. In contrast to this, some other studies conclude that India does not require a public agency of the size and type that FCI is today and that there are several negative effects of government of India's foodgrain marketing policy and operation of FCI which are found to be increasingly costly and inefficient. Technical and managerial inefficiencies in FCI operations are said to be responsible for its high costs.

Though one can find some justification in economic cost of FCI but this justification does not take into account implicit value of quality deterioration of produce at various levels. This happens due to purchase of lower than specified grade of produce, weight manipulations at points of purchase and dispatch, excessive charges of FCI contractors, and adulteration and supply of poor stuff under levy and custom milling of rice. Deterioration in value of produce resulting from such practices is the main source of leakage in FCI operations and is not reflected in cost or price calculations. The produce get sold because it is offered to various states at a subsidised price and the difference between economic cost and issue price is shown as food subsidy. This way the inefficiency of FCI is concealed.

One way to ascertain and quantify this kind of leakage is by comparing market value of PDS supply with ruling market price of FAQ produce. If PDS supply can sell at the same rate as the ruling market price then there is no quality deterioration. If it sells at a price lower than market price, the difference represents allowance for quality deterioration or leakage. The studies that justify cost of FCI, do not reckon this aspect. If such leakage are accounted for separately, food subsidy bill would go down significantly and we feel it is not difficult to check such leakage.

We collected some samples of rice and wheat supplied through PDS outlets from Orissa, Andhra Pradesh and Rajasthan to compare the value of PDS produce with open market price of produce having same quality as PDS supply and with economic cost of produce to FCI.

In Rajasthan wheat was sold to below poverty line households at the issue price of Rs. 4.60/kg. We took the sample of PDS supply to find out its market price. The price of produce of same quality as sold in PDS varied between Rs. 5.50 to Rs. 7.00 in open market. Thus market value of PDS supply was below the economic cost of FCI which was Rs. 8.71 for the year 2000-02.

The market value of PDS rice in Orissa and Andhra Pradesh was quite low compared to economic cost of FCI. The market value was found to vary between Rs. 6.00 and Rs. 8.00 whereas the economic cost of rice to FCI was Rs. 12.04. The actual subsidy going to PDS consumer is the difference between issue price and market price of the same quality of produce. Whereas, as per FCI account books, the subsidy accruing to PDS consumer is taken as the difference between economic cost and issue price. The difference in market price and economic cost of FCI mainly results from quality deterioration of stock with FCI, which is not reckoned in calculation of economic cost.

The inefficiency and high cost of FCI are often used to make case for binding up FCI and to pave the way for greater private sector participation. In this context it is worth mentioning that in the absence of public agencies, private trade may turn out to be exploitative and what now go as inefficiency of FCI would go as excessive profit of private trade. Therefore this public agency should be retained but it should plan its operations in such a way so as to keep check on private trade to exploit market situations. However, the area of operation of this parastatal should be reduced and its efficiency should be improved by modernisation of its operations on scientific lines and by professionalisation of its management.

Suggestions for Government Intervention in Food Management

The comfortable situation on foodgrain front, new options like export-import to stabilise supply, growing burden of cost of food management, production imbalances caused by food policy, strong regional bias and some other problems associated with food management provide strong reasons to look into the policy of food management and change it appropriately.

This would involve several steps beginning with guaranteed support prices. The purpose and the basis of these prices need to be reviewed periodically with the change in circumstances. These prices must reflect the society's preference for the produce and should promote efficiency and quality. In the present form the guaranteed prices have given rise to several problems.

As it is not feasible to ensure that prices would not fall below MSP in any commodity, only selected crops should be covered under MSP. It is suggested that crops which can be considered as price leader or the crops for which technological breakthrough is imminent ought to be covered under the MSP and other candidates for support price would be the crops grown in high risk environment. In the long run country needs to develop new mechanisms to provide protection to farmers income. Achieving this objective through price intervention alone results in several distortions. Government should provide support to develop viable crop insurance for protecting crop income.

Due to changes taking place in consumption basket of food, there is lot of emphasis to develop technologies that promote diversification of agriculture sector. Price interventions should be such as to encourage agricultural diversification.

There is also a need to discuss criteria on which MSP should be based in the changing context. The popular perception is that MSP is determined based on cost of production. When the emphasis of production is shifting from food security to market led production, it is not justified to base MSP on cost of production. Similarly, there are concerns relating to definition of cost of production on which MSP should be based. Some of the cost concepts like Cost C3 are such that the price based on those is not quite relevant to qualify as "minimum support price".

Second, government intervention should be such that it promotes private trade and competition. With the development of countrywide transport and communication network, availability of different items has improved considerably in all parts of the country. Therefore, if incentives are favourable, private trade should be able to supply grains everywhere including remote areas.

Government intervention in the form of procurement should be selective. In a normal production year, quantity of procurement should not exceed PDS requirement. There is a need to maintain food security buffer stock which should be maintained by purchasing grains during above normal production and releasing stock during low harvest years. The level of buffer stock around 7 million tonne would be adequate to meet supply shortfalls in most of the years.

If the level of minimum support price reasonably corresponds to the price level as dictated by demand and supply factors than market price should not go below MSP in competitive markets in normal year of production. Hence the need for government procurement in normal years of production should not arise. When production is forecast to be above normal and this information is available at harvest time then government should procure above trend quantity. This should ensure that prices do not go below MSP. Still there could be cases where private trade turns out to be exploitative and farmers are paid price below MSP. One way to address this kind of situation is to make "deficiency price payment" to farmers covering a part of the difference between actual price received by farmers and MSP. Similarly, it is not possible to carry out procurement in all the markets in the country to ensure MSP, and, stock position may not justify procurement in some years. The system of deficiency price payment can work as an alternative to procurement operations in such situations. This would help in preventing unwanted stocks and help in providing price incentive to producers in all the regions considered relevant for the purpose. Thus the system of "deficiency price payment" can help in achieving economy in procurement and regional equity in implementing guaranteed price.

The most serious problem with "deficiency price payment (DPP)" is said to be its administration and problem due to recycling of produce. For this we suggest that the system of deficiency price payment should be implemented for the produce sold through regulated markets in all surplus generating areas using district as a unit for determining surplus area. In order to ensure that resale of produce does not take place, the magnitude of deficiency payment should be kept less than the charges involved in first sale of produce like mandi fee, auction, labour charges etc. This kind of mechanism would not suffer from problems like regional bias and excessive stocks.

The amount of foodgrains needed for PDS supply and inter year price stabilisation should be purchased through competitive bid from the markets where prices are ruling lowest.

Policy Recommendations

India's food management system has undergone tremendous changes since the launch of green revolution technology during late 1960s. The most profound impact has been witnessed in the case of rice and wheat whose production has shown spectacular increase and enabled the country to move from chronic food shortage to an era of food surplus. Besides technology and increased use of inputs, government intervention and support in terms of guaranteed minimum prices for output and large scale procurements from producers at remunerative prices have played key role in growth of output of rice and wheat. These government interventions have been affected through price policy and parastatls like Food Corporation of India (FCI) which procures rice and wheat from surplus regions, maintain buffer stock, supplies rice and wheat for public distribution system in various states and undertakes open market sales. This intervention has been quite effective in achieving goal of food security for the country but its sustenance has been putting large burden on the state exchequer in the form of food subsidy which has been growing year after year. Government intervention is also found to be strongly biased towards some regions and has caused adverse impact on quality and efficiency of production and participation of private trade in grain marketing.

Demand and supply scenario for agriculture has undergone profound changes during last 10-15 years. However, farm price policy and policy for food management have not been changed to adjust to new situation. The result has been that Indian agriculture in the beginning of 21st century is facing some serious contradictions and challenges. On one hand the country has 60 million ton of foodgrains in public stock and on the other hand every fourth Indian is reported to be underfed and does not get even minimum calorie required by human body. Similarly, the country is facing massive shortage of pulses and edible oils, which is met through imports, having detrimental effect on the producers in the unfavourable dryland areas, whereas, surplus of wheat and rice is being sold as export at a large cost to the state exchequer. These imbalances are causing serious concern to the country.

The reason for emergence of demand and supply imbalances in Indian agriculture lies in both, technological and policy factors. The technological breakthrough of green revolution category has been highly biased towards cereals, particularly wheat and rice, which raised their relative profitability and has led to large shift in land and other resources away from other crops. The technological advantage has been further provided strong policy support through remunerative and assured prices for rice and wheat. The stage has now been reached where current level of rice and wheat production cannot be absorbed at existing level of their prices while there is massive shortage of edible oils and pulses.

One of the reasons for continuing high policy support to grains till recently is that demand projections for grains did not take note of diversification in consumption pattern experienced in rural as well as urban areas. Because of changes in food habits, demand for non cereal foods is likely to rise by more than 50 percent of the growth in cereal demand. Similarly, growth in supply of major commodities like edible oil and pulses has not been keeping pace with the growth in demand. This has resulted in fast growth in import of edible oils while pulse deficit is reflected in both imports as well as in the increase in domestic prices of pulses. There is a need for suitable policy to address these imbalances.

Under liberalised trade, export and import are considered the main instrument for stabilisation of domestic demand and supply. However, due to high volatility in international prices and because of being residual exporter, India could not use trade options very satisfactorily to stabilise supply of foodgrains. The trade option has particularly become difficult to dispose off large stock of rice and wheat in India in international market in the recent years. International prices of grains have dropped sharply even in Rupee terms after 1998-99 but domestic prices have been moving up mainly under the pressure of increasing support prices. This has reduced competitiveness and profitability of grain exports. India is able to sell its surplus stock in foreign market at a huge subsidy in relation to economic cost and open market prices. Wheat and rice are being released for export at a price charged from below poverty line households for PDS supply.

In India about 26 percent population is reported to be below poverty line and is undernourished. Cereals are the main source of calorie and food security for such population. In the recent years huge quantity of cereals (rice and wheat) has piled up in government stocks which constitutes more than one third of the total production of rice and wheat in the country and is far above the quantity genuinely needed in the buffer stock. This has reduced availability of grains in the country, implications of which need to be understood properly.

On production side, per capita production as well as availability of cereals, have moved on a somewhat rising trend during 1970 and 1980s. Per capita production of cereals reached peak level around 1990, after which cereal output and India's population have risen almost at the same rate. Per capita availability, which in a somewhat crude sense can be considered to represent per capita consumption, kept pace with trend in per capita production of cereals during 1970s and 1980s. However, after 1990 per capita consumption of cereals has shown a decline. This way, post 1990 period show two clear indications relating to cereal production and consumption in India. One, growth rate in cereal production has decelerated to the level of population growth. Two, average availability of cereals has followed a decline in the recent years and it has gone down to the same level as observed during 1987 which was a very bad drought year.

The recent decline in per capita availability of cereals is neither a result of slack in production, nor it has resulted from export of cereals. This is entirely due to accretion to stock of cereals held by government agencies. There are two reasons for production going to stock rather than being consumed. One, PDS prices during 1990s have increased in jumps and at a faster rate compared to open market prices and prices of other food and other commodities, which caused a decline in per capita PDS demand in the recent years. Similarly, open market retail prices of rice and wheat have also increased at a much faster rate during 1990s compared to 1980s. Thirdly, consumption pattern in India is changing towards non cereal foods.

The change in consumption basket is being considered a healthy trend in some quarters whereas some researchers consider it adverse for the poor. The decline is explained due to increasing mechnisation of human activities, reduced physical exertion and improved access to modern transport facility and income effect on food preference for variety in food intake and for nutritious and healthy food rather than increasing calorie intake. On the other hand, it is contended that improvement in per capita income should result in fast increase in the indirect demand for cereals which should lead to increase in overall per capita demand for grains in the county. The argument is based on the logic that improvement in income leads to fast growth in demand for livestock products, which should result in higher growth in the use of food grain as feed as experienced in countries like China. Here, there is a need to distinguish Indian situation from other countries. Despite observed growth in demand for livestock products, per capita consumption of meat, egg and fish in India remains very low. This seems to be largely due to cultural factors. Even if it can be afforded, people in India do not eat meat etc. on a regular basis. For a vast majority of those who are non-vegetarian, meat intake is preferred occasionally and not as a part of regular diet. Therefore, India is not likely to witness serious competition between food and feed uses of grain. Similarly, according to Indian value system, there is emphasis on eating only that much which is needed for survival and there is no tendency to eat more and expend the same by doing exercise etc., unlike in developed countries. In the light of these factors, cereal demand in India should not be expected to grow at faster rate than the population growth rate.

Rice and wheat are the staple food for Indian population. Therefore, their prices cause significant impact on the well being of general population. These two commodities being wage goods, determine consumption level of vast majority of population and have strong effect on overall price level of the economy. Stability in prices of these two commodities has remained an important goal for the country. One of the obvious factors affecting price level and changes therein is production level of the commodities. However, it has been often observed that even during the years of normal production sometime prices tend to show abnormal rise.

In the case of wheat, annual change in its output, buffer stock of wheat as on 1st July, export of wheat during the year, change in procurement price over the previous year, change in issue price of wheat during the year and per capita offtake of wheat for PDS were found to be the main factors which explain variation in monthly and yearly prices of wheat. Issue price of wheat caused stronger impact on open market prices as compared to the impact of procurement price. Increase in issue price raised open market monthly price to the same magnitude as the percent change in issue price.

Unlike wheat, policy variables were less relevant in affecting changes in prices of rice. Importance of different variables also differ for intra year and inter year variation. Intra year price variation was caused mainly by changes in procurement price, quantity supplied through PDS, and rice export. Inter year price variation resulted mainly from changes in output of rice, output of other cereals, procurement price and issue price. Variation in rice output caused much smaller variation in prices compared to that in wheat. Similarly, rice price was relatively less sensitive to export compared to wheat. The reason for different policy variables not showing similar impact on rice price, unlike wheat, could be lesser seasonality in rice production and wider production base of rice.

As intra and inter year price variations are caused by several factors and their interacting influences, there are instances when prices of rice and wheat had risen very high despite normal or even above normal domestic production. This can happen when combined effect of other variables dominates the effect of production on prices.

When India initiated economic reforms in year 1991, a move towards increased participation by private sector in various spheres of economic activities was made. In the case of agriculture, particularly the food grain trade, some reduction in the role of government took place between 1992-93 and 1996-97. However, this decline turned out to be the part of cyclical variation resulting from production fluctuation as experienced in the past and government procurement witnessed steep increase in the recent years.

The difference in the current and the past situation is that government procurement increased despite very comfortable situation of food stock and below normal production. The reason has been the reduced role of private trade in grain trade. Huge buffer stock with government created a fear with private trade that release of excessive stock could dampen the domestic prices at any time.

The government's procurement, distribution, and buffer stocking programmes are reported to have had negative impact of repressing private foodgrain marketing, undercutting its potential contribution to long term food security. This is further said to discourage modernisation of marketing resulting into losses and inefficiencies. It is being proposed that government should use regulatory mechanism only when price movements are outside the desired price band representing width between the ceiling and floor price, which permits reasonable marketing margin for profitable public sector operations. This underscores the need to evolve new kind of mechanism for government intervention in food grains markets which allows and encourages active participation of private trade but keeps a check on exploitative tendencies of private trade. This can be done by creating an environment that provides reasonable incentive to private sector to operate in food grain market. This implies that as long as private trade charges normal margin for delivering the commodities across space and time, government would not intervene in the market. Implementing such mechanism would require developing norms which can be used to find out whether private trade is charging reasonable prices or is indulging in exploitation of consumers and producers. We have prepared estimates of price band between farm harvest prices of wheat and paddy in surplus states and wholesale and retail price in subsequent months in all major states of the country. The acceptable price band includes reasonable margin for wholesalers, rice millers and retailers, and, marketing costs like (1) statutory charges, (2) labour charges, (3) material (packaging) cost, (4) local transport and freight to different states (5) milling charges for rice (6) storage charges and (7) interest on working capital.

At existing structure of statutory charges/taxes, and transport and other costs, retail price for wheat in surplus states should be higher than farm harvest price by 36 to 60 percent in various months to attract private trade.

In the deficit state like Maharashtra, retail price before harvest need to be more than double the farm harvest price in surplus states to provide reasonable incentive to private trade. Similarly the band suggests that retail prices of rice should be 96 to 213 percent higher than the farm harvest price of paddy to attract private trade to buy paddy and supply rice in various months and in various states. Price difference beyond this band would imply exploitation by private trade which would need government intervention. Transport cost and statutory charges are the main element of price spread and price band can be narrowed down by curtailing these costs.

Due to inelastic nature of their demand and supply, small fluctuation in output of cereal is likely to cause large increases/decreases in their prices which in turn would result in sharp fluctuations in farm income and expose consumers to high volatility in food prices. Thus, in order to bring stability in prices and farm income it was considered imperative to maintain buffer stock of grains which involved purchases from the market during the good harvest years and releasing stock during lean years or when production is below envisaged trend. Purchase from market and release of stock in this manner ensures that supply of produce in the market would move on a smooth trend. If growth in demand keeps pace with the envisaged trend in supply then purchase/release of stock based on deviations from trend level of production would ensure perfect balance between demand and supply assuming closed market from external trade. This should ensure that there is no disturbance to price stability as for as genuine factors on demand and supply side are concerned.

Based on this logic, deviations in actual production from stipulated trend should serve as a basis for quantity of grain to be purchased and buffer stock needed for maintaining inter year price stability. During the decade of 1990s fluctuations in cereal output declined considerably as deviations from normal production remained below 7.06 million tonne on either side. However, procurement by official agencies during 1990s rose sharply. Like the increase in procurement of foodgrains by government, level of buffer stock of cereals has also moved up over time. Thus, reduction in instability in cereal production in the country was not accompanied by decline in quantity of output purchased by government and buffer stock. Quantity of foodgrains purchased by government and level of buffer stock have not been dictated or guided by fluctuations in output. If maintaining inter year price stability was the important consideration for government intervention in grain markets then reduced fluctuations in output, experienced in the country, should have resulted in decrease in level of procurement and buffer stock rather than following increase over time.

Assuming that private trade would not play major role in maintaining inter year price stability, government can base its decision to buy above normal output on advance estimates of production available around harvest time. This would be given by deviation of estimated output from stipulated trend, which can be estimated quite reliably for medium term. Similarly, level of buffer stock for such purpose can be worked out from negative deviations of actual output from stipulated trend which indicate that buffer stock of around 7.63 million tonne is sufficient to make up for the decline in cereal output from the normal level.

There would be some instances of severe drought, like those experienced during 1979-80 and 1987-88 and recently during 2002, when decline in output could go quite high. This kind of event is highly unpredictable about its severity as well as occurrence. It would not be prudent to link level of buffer stocks to magnitude of output decline in such years as it requires maintaining huge level of buffer stock for a very long period. Normal buffer stock would take care of about half of the output shortfall in the time of severe draughts. Some of the output deficiency in such events can be met from international market, and, some downward adjustment in consumption can also compensate a part of it.

Total requirement for PDS is estimated to be 21 million tonne per annum. Procurement for PDS would be a continuous exercise requiring purchase of rice and wheat each year. Thus, total procurement of grain in a given year should be around the sum of quantity needed for PDS and excess of estimated output from normal production. Similarly, maximum level of buffer stock should be 18.13 million tonne, which is sum of stock needed for PDS for six months and highest shortfall observed in grain production during the recent 10 years.

Government ensures guaranteed prices to producers by buying grains delivered at the support prices. If support price is consistent with demand and supply for the given commodity then competitive market should ensure that market price does not go below the level of support price. In such a situation there would be no need for government procurement to ensure minimum price to producers. However, if market is not competitive or is unfavourable for private trade then there could be need for government to ensure minimum support through purchases from the market.

In case market price determined by supply and demand does not leave adequate margin over support price, private trade would not buy the produce at support price, howsoever competitive the market may be. This can happen when growth in demand does not keep pace with the supply and, support price is fixed based on supply side factors. This is exactly what is being experienced in the case of cereals in the country.

Indian consumers meet their demand for cereals from own production, open market purchases and PDS supplies. Thus, open market prices at retail level and PDS prices matter in determining demand. Both PDS price as well as retail prices have increased sharply during 1990s leaving adverse effect on demand. Demand for cereals has also been hit due to changes in consumption pattern away from cereals towards non cereal food.

Stagnation in per capita production and rising stock with government in the recent years indicate that government intervention in rice and wheat market through prices, procurement and buffer stock has reached a stage where it is depriving common consumer of availability of basic food.

To the extent deficiency of cereal demand is resulting from long run changes in consumption pattern, prices would not be very potent instrument for boosting demand and for restoring balance between demand and supply. The solution lies in adjusting supply to demand.

If demand for cereal remains deficit and cereal output increase to their normal (trend) level than required decrease in price to maintain balance between supply and demand works out to be enormous. This kind of purely price based solutions would result in steep decline in crop income and would cause adverse impact on agriculture sector which is already under threat due to trade liberalisation, and, is not able to cope up with low level of international prices.

The best way to ensure remunerative prices to foodgrain producers and reducing unwanted surplus in future is to take measures that result in some shift in resources from cereals to non cereal enterprises and encourage growth of cereal output in efficient producing regions. This strategy should not be based only on reducing profitability of grain production by lowering their prices but it should involve providing and developing alternatives which are more remunerative than cereals.

Experience shows that mere announcement of higher support prices for commodities, which are not effectively backed up by procurement arrangement, does not serve the purpose of raising level of prices received by producers. Therefore, attempts by CACP to raise support prices of crops like edible oils and pulses, in which India is deficit, relative to support prices for rice and wheat which are in excess supply, cannot be expected to result in shift of resources from rice and wheat to the deficit crops.

Another serious criticism of government intervention in grain markets is that regional concentration in government procurement of grain has remained quite strong. Proportion of marketed surplus procured by official agencies across states vary from below 2 percent to more than 85 percent. During recent years there have been frequent reports from states of Orissa, Madhya Pradesh, Bihar about distress sale of rice and maize below MSP. These states have been late adopter of new technology. Though they are food deficit at aggregate state level, several growth pockets have emerged in these states having surplus foodgrains. These pockets are in the first stage of green revolution and agricultural development, when private trade and market institutions are not in place to provide incentive to encourage adoption of new technology and hence output growth. Agriculture growth would get a serious setback in such areas if institutional support in the form of guaranteed price is not provided.

Several reasons are responsible for accumulation of grain stock much above the genuinely required level. First, during the decade of 1990s procurement prices of rice and wheat were given a comfortable increase which was quite higher than the increase in general prices. This was followed by steep rise in procurement by government agencies in surplus regions. The increase in procurement prices necessitated increase in issue prices and, because substantial part of marketed surplus was with the government, its policy on quantity and price of stock released had a determining influence on open market wholesale prices of rice and wheat. The influence on wholesale prices was such that they increased by even higher rate then procurement prices. As demand side factors did not support this increase in price, growth in retail prices started trailing behind the growth rate in largely government influenced wholesale prices. This caused adverse impact on the margin of private trade which slowly started withdrawing from the market. Accumulation of cereal stock with government agencies created a feeling that release of excess stock can anytime depress open market prices. These two factors led to withdrawal of private trade from grain markets in surplus states causing increase in procurement by government agencies, even when there was no shortfall in production.

Second, increase in PDS prices and diversification in food consumption habits led to demand deficiency resulting in diversion of production to inventory. Third, steep decline in international prices of cereals since 1997 has caused adverse impact on export of cereals and excess stock could not be smoothly sold in international market.

Unless excessive stocks are liquidated, grain markets in India would continue to remain in trouble. There is no single and easy way to do this and several options need to be tried. Some of such options are (i) increasing BPL quota (ii) expanding employment generation programmes (iii) universal PDS at BPL prices in calamity affected areas. In addition to these, some other options also need to be tried. Very old stock that has deteriorated in quality should be disposed off as a feed in international and domestic market at whatever price it can fetch. If still excessive stock remains, some of it should be given as food aid to needy countries.

Buffer stocks have been used by the government as an important instrument for the purpose of price stabilisation. However this involved heavy cost in terms of procurements, handling, carrying, storage etc. which is becoming fiscally unsustainable. As an alternative it has been suggested that government should use the instrument of variable levies on external trade to stabilise domestic prices. A comparison of domestic stabiliation measures and trade shows that selling and buying wheat in international market to stabilise domestic output does not result in large changes in international prices of wheat due to large volume of world trade in wheat. However, in the case of rice, stabilising domestic supply through trade caused sharp fluctuation in international price of rice. Among the two options viz. domestic stabilisation through buffer stock and stablisation through trade the latter is found to be costlier than domestic stabilisation in most of the years though it also depend upon fluctuation in international price. If the relationship between domestic and international prices in future remains same as observed during the last 26 years, than policy of price stabilisation through buffer stock seems to be better option than trade.

Food Corporation of India has remained in the centre stage of government intervention in agricultural marketing due to scale of its operation and due to its role in food security. Though this agency has played significant role in ensuring guaranteed price, and hence in adoption of improved technology in traditional green revolution region, its cost of operation and efficiency have remained subject of criticism and are seen as the main factor for mounting food subsidy in the country. The economic cost of FCI that is often used to justify its operation, does not take into account implicit value of quality deterioration of produce at various levels. This happens due to purchase of lower than specified grade of produce, weight manipulations at points of purchase and dispatch, excessive charges of FCI contractors, and adulteration and supply of poor stuff under levy and custom milling of rice. Deterioration in value of produce resulting from such practices is the main source of leakage in FCI operations and is not reflected in cost or price calculations. The produce get sold because it is offered to various states at a subsidised price and the difference between economic cost and issue price is shown as food subsidy. This way the inefficiency of FCI is concealed.

The inefficiency and high cost of FCI are often used to make case for binding up FCI and to pave the way for greater private sector participation.

In this context it is worth mentioning that in the absence of public agencies, private trade may turn out to be exploitative and what now go as inefficiency of FCI would go as excessive profit of private trade. Therefore this public agency should be retained but it should plan its operations in such a way so as to keep check on private trade to exploit market situations. However, the area of operation of this parastatal should be reduced and its efficiency should be improved by modernisation of its operations on scientific lines and by professionalisation of its management.

Minimum support prices for various commodities must reflect the society's preference for the produce and should promote efficiency and quality. In the present form the guaranteed prices have given rise to several problems. As it is not feasible to ensure that prices would not fall below MSP in any commodity, only selected crops should be covered under MSP.

In the long run country needs to develop new mechanism to provide protection to farmers income. Achieving this objective through price intervention alone results in several distortions. Government should provide support to develop viable crop insurance for protecting crop income.

Due to changes taking place in consumption basket of food, there is lot of emphasis to develop technologies that promote diversification of agriculture sector. Price interventions should be such as to encourage agricultural diversification to address imbalances in Indian agriculture.

There are concerns relating to definition of cost of production on which MSP should be based. Some of the cost concepts like Cost C3 are such that the price based on those is not quite relevant to qualify as "minimum support price". There is a need to develop more relevant norms/ criteria for price support.

Still there could be cases where private trade turns out to be exploitative and farmers are paid price below MSP. One way to address this kind of situation is to make "deficiency price payment" to farmers covering a part of the difference between actual price received by farmers and MSP. Similarly, it is not possible to carry out procurement in all the markets in the country to ensure MSP, and, stock position may not justify procurement in some years. The system of deficiency price payment can work as an alternative to procurement operations in such situations. This would help in preventing unwanted stocks and help in providing price incentive to producers in all the regions considered relevant for the purpose. Thus the system of "deficiency price payment" can help in achieving economy in procurement and regional equity in providing incentives to producers.

The system of deficiency price payment should be implemented for the produce sold through regulated markets in all surplus generating areas using district as a unit for determining surplus area. In order to ensure that resale of produce does not take place, the magnitude of deficiency payment should be kept less than the charges involved in first sale of produce like mandi fee, auction, labour charges etc. This kind of mechanism would not suffer from problems like regional bias and excessive stocks.

Government intervention should be such that it promotes private trade and competition. With the development of countrywide transport and communication network, availability of different items has improved considerably in all parts of the country. Therefore, if incentives are favourable, private trade should be able to supply grains everywhere including remote areas.

Government intervention in the form of procurement should be selective. In a normal production year, quantity of procurement should not exceed PDS requirement. There is a need to maintain food security buffer stock which should be maintained by purchasing grains during above normal production and releasing stock during low harvest years. The level of buffer stock around 7 million tonne would be adequate to meet supply shortfalls in most of the years. The amount of foodgrains needed for PDS supply and inter year price stabilisation should be purchased through competitive bid from the markets where prices are lowest.

References

Chand Ramesh. *Trade Liberalisation, WTO and Indian Agriculture: Experience and Prospects*, Mittal Publications, New Delhi.

Jha Shikha and P.V. Srinivasan. Food Security through Price stabilisation: Buffer Stocks vs Variable Levies, *Economic and Political Weekly*, 34(46 &47): 3299-3304, November 20.

Johl, S. S.. Agriculture Sector and New Economic Policy, *Indian Journal of Agricultural Economics*, Conference no. 50(3): July -September.

Srinivasan P.V. and Shikha Jha. Food Security through Price stabilisation: Buffer Stocks vs Variable Levies, *Economic and Political Weekly*, 34(46 &47): 3299-3304, November 20.

Swaminathan Madhura. Understanding the Costs of Food Corporation of India, *Economic and Political Weekly*, 34(52): A121-AA132, December 25.

7

World Rice Market: Emerging Issues and Trends

Rice is the most important food crop with more than 90 percent of global production occurring in tropical and semi-tropical Asia. In several Asian countries, rice provides between 50 and 70 percent of the energy and protein dietary requirements. Rice is also a staple food in many Latin American and Caribbean countries, with consumption levels among the low-income strata approaching the levels witnessed in Asia. Rice is the most rapidly growing food source in several African countries and Nigeria has emerged as a major importer of rice. The rapid acceleration of rice production over the last three decades has been a primary contributor to improvements in world food security. However, there are still 800 million people suffering from food deficits, most of them residing in areas that are dependent upon rice production for food, income and employment.

Since the Nineteenth Session of the International Rice Commission (IRC), held in Cairo, Egypt in 1998, there have been numerous developments having a profound impact on the global rice industry. Rice production has continued to expand and exceed consumption, while several countries have begun dispersing the huge rice stocks accumulated by bringing down rice prices in international markets. In several countries, increased incomes have led to a downward trend in per caput rice consumption. Low rice prices are a concern for rice-exporting and rice-importing countries which also have domestic production, as the latter could be threatened by less expensive imported rice. The depressed economic situation of rice and significant government support to maintain national production and rural incomes may have serious consequences for World Trade Organization members and their commitment under the Uruguay Round Agreement on Agriculture.

With the widespread adoption of hybrid rice, China witnessed production increases, while there was a decrease in the area under rice during the early 1990s. Biotechnological tools for manipulating rice genetics are still in an early development stage but hold promise for addressing several constraints in plant breeding; indeed, improved techniques and the increase in genomic information have resulted in great advances in rice biotechnology. There have been positive developments, but polarization between the opponents and supporters of transgenic food crops has intensified, representing a serious issue for future access to and adoption of genetically transformed rice varieties.

This paper examines the current situation of rice production with the objective of highlighting issues and concerns related to economically sustainable production. The current rice situation and long-term demands are far from separate issues, since remedies to the current crisis could have a major influence on meeting future needs.

More than half the world's population depends on rice for its major daily source of food energy and protein, and four-fifths of the world's rice is produced and consumed by small-scale farmers in low-income developing countries. Two billion people in Asia alone derive 80 percent of their food energy intake from rice, and thus the importance of this crop in relation to food security and socio-economic stability is self-evident. In many countries rice accompanies every meal, and it is an integral part of religious ceremonies, festivals and holidays. In high-income countries in the Near East, Europe and North America, it is becoming increasingly popular because it is seen as a healthy and tasty food.

The increase in world rice production during the green revolution resulted in more rice being available for consumption despite the continually growing population. During the past four decades, the unit cost of rice production decreased by 30 percent while the price of rice fell by over 40 percent, which made an important contribution to lessening hunger and poverty and enhancing and sustaining livelihoods. This led to a belief in the success of the 1960s Freedom-From-Hunger-Campaign. However, there are about 800 million people in the world today suffering from malnutrition. Hunger reduces people's potential to work, hinders children's learning capacity and leads to a vicious cycle of poverty. The theme of the Twentieth Session, therefore, is "Rice for Food Security".

The security and sustainability of rice-based livelihoods is fundamental to the world's food security. The World Food Summit (WFS), convened in November 1996 in Rome, called for coordinated global action to secure food security for the world's population. The WFS Five Years Later, held from 10 to 13 June 2002 in Rome, renewed the global commitment made in the Rome Declaration of the first Summit to accelerate the implementation of the WFS Plan of Action. Increasing rice production and distribution would undoubtedly contribute to the successful implementation of the WFS Plan of Action.

However, there are increasing concerns about the ability of rice production to meet popular demand in the near future. The Expert Consultation on "Technological Evolution and Impact for Sustainable Rice Production in Asia and the Pacific", held in Bangkok in 1996, reported the stagnation of rice yield in many Asian countries. Furthermore, the intensification of rice production, if not pursued scientifically and appropriately, has been responsible for considerable damage to the environment and natural resources, including the build-up of salinity and alkalinity, as well as for water pollution and health hazards resulting from the excessive use of agrochemicals. In the past, the contribution of lowland rice to greenhouse gas emissions has been unjustifiably overstated. On the other hand, concern about genetic erosion in rice production is increasing.

Fortunately, technological options for sustainable rice production are available. Further yield increases may be achieved with new generations of varieties, including hybrid rice, new super plant types and transgenic rice. Innovations in rice integrated crop management (RICM) could improve the efficiency of input utilization and reduce the risk of environmental pollution and production costs. Labour-productivity enhancement is expected to be achieved with new generations of agricultural machinery, whilst innovative harvest and post-harvest technologies can reduce postharvest losses. Increased accessibility to information concerning innovation and production factors through modern communication systems can expedite the adoption of new technologies. New successful experiences in bridging the rice yield gap - for example, the "Ricecheck" approach - if widely adopted, could significantly enhance rice productivity and production.

The prevailing decline in the world's rice prices due to bumper crops during the last few years should not lead to mistaken complacency. On the one hand, the declining price of this vital commodity has greatly helped to alleviate poverty and hunger in many cities in the world, but on the other hand it has caused hardship to rice producers. In this context, as globalization progresses, the world community needs to provide safety nets for the world's vulnerable rice smallholders.

Moreover, world rice production suffers from lack of investment in irrigation development and research work. The decline in the development of irrigation infrastructures has slowed down the adoption of high-yielding varieties and improved crop management techniques. Similarly, the substantial decrease in public investment in rice research is troublesome, since the difficulty of sustaining growth in rice productivity has increased as yield has advanced. Water will be the scarcest resource in the years ahead, and the water-rice relationship will occupy centre stage in enhancing and sustaining the rice-based system.

In view of the above, the challenges faced by today's national and international rice research and development programmes are twofold:

1. How to render rice production technologies both efficient and compatible with the limited resources of poor farmers, while being environmentally friendly and sustainable.
2. How to design rice and rice-based production systems that fit into the socio-economic environments (especially where the participation of women is dominant) and that are capable of reducing the risks of rural poverty.

Trends of Supply and Demand

Figure 1 presents information on global rice production, consumption and ending stocks for the period from 1997 to 2001/02. These data illustrate the two distinct phases that have transpired over the last 5 years. From 1997 to 1999, production exceeded consumption; overproduction was mainly a result of a combination of favourable international prices in previous years, support prices and government support to export in several producing countries. Following these "surplus years", global rice production has declined during the last 2 years; consumption has outpaced production. Global stocks have also decreased to compensate for the decline in production. However, global stocks in the last 2 years have remained high. The current production trend is expected to continue until rice stocks reach a more realistic level and demand and supply factors cause international rice prices to rise.

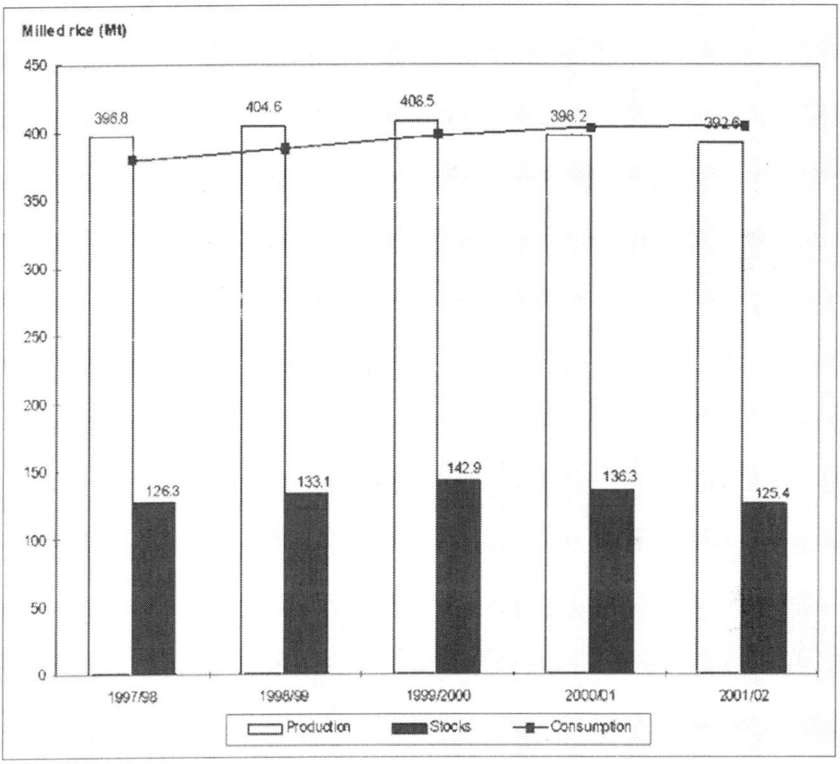

Figure 1. Worldwide trends in rice production, consumption and ending stocks

During the early 1990s, several scientists warned of a pending crisis. The rice production growth rate fell from over 2.5 percent during the 1980s to only 1.1 percent per annum during the early 1990s as a result of slow yield growth and limited expansion of the production area. Several reports stated that a 75 percent increase in production will be required by 2025 to meet the demand for approximately 850 million tonnes (Mt) of paddy. However, a more recent analysis points to lower demand for rice in both the short term (2010) and the long term (2030). These projections are based upon recent consumption trends, taking into account decreasing per caput rice consumption due to urbanization and increased income. Between 2002 and 2010, rice consumption in China is projected to decrease at a rate of approximately 0.45 percent per year, while in other Asian countries, consumption is forecast to remain stagnant. The same scenario is projected through to 2030 (Table 1). The forecast of economic growth of 4 to 5 percent per year is the driving force behind the anticipated reduction in rice consumption.

Table 1.Projections of world rice demand for human food in 2015 and 2030 and growth rates in demand

	Rice demand (milled) (Mt)			*Per caput growth in demand (%/year)*		
	1997/99	2015	2030	1999	2015	2030
World	386	472	533	1.2	0.8	1.0
Rice as food (kg per caput)						
East Asia	106	100	96			
East Asia, excluding China	132	129	124			
South Asia	79	84	81			

Note: Projections take into account current estimates of GDP growth, emerging consumption patterns, population trends and other variables, such as urbanization.

In summary, rice demand in 2030 is projected to be approximately 533 Mt of milled rice. While significantly lower than earlier projections, a considerable quantity of rice will still be required to meet future needs. Moreover, although global rice projections do not anticipate deficits, they often mask regional or national requirements. Indonesia is expected to continue to experience rice deficits increasing from the current annual level of 3.6 Mt to over 4.4 Mt in 2010. At present, Nigeria has annual imports of about 1 Mt, projected to grow to 1.8 Mt in 2010. Sub-Saharan Africa is projected to import over 6 Mt per year by 2010, while Brazil, Cuba and Mexico are forecast to experience continuing rice deficits of approximately 1.5 Mt per year.

Pertinent Issues

Several emerging issues may hinder the ability to meet future rice needs. Yield stagnation and limited land and water resources for expanding rice area are the most important constraints to sustainable rice production.

Concerns for nutritional quality, genetic erosion and environmental degradation require more stringent choices in rice production, especially in relation to international commitments.

Yield Stagnation and Productivity Decline

The growth in rice yield has decreased to slightly more than 1 percent per year, i.e. approximately equal to population growth. Yield stagnation appears to be a problem even with hybrid rice. There is extensive evidence of declining yields and falling productivity in the 28 million ha (Mha) of irrigated rice under intensive cultivation in Asia. Reports from the early 1980s showed a yield decline in the intensively cultivated rice plots in research stations in the Philippines. Later studies in other countries also reported yield declines. More recent analyses also support the earlier reports of declining yields in research trials; however, they are unable to ascertain the extent of the decline in other parts of Asia, due to year-to-year variation in yields. Other observations indicate that continuous cultivation of irrigated rice (where the soil is maintained in anaerobic conditions for prolonged periods) results in disorders that limit yield. Paramount among the constraints is the production of toxins from the anaerobic decomposition of organic matter that may limit plant development and subsequently yield. Given the importance of the intensive production system, attention must be directed towards sustaining high yields and avoiding inducing yield-limiting factors.

Genetic Uniformity and Erosion

It has long been known that irrigated rice production is highly vulnerable to major biological attacks as a result of genetic uniformity. In the 1980s, IR 36 was grown on approximately 13 Mha in Asia. Currently, IR 64 occupies 10 Mha - about 15 percent of all irrigated riceland. A similar situation exists for hybrid rice. In 1998, it was reported that 95 percent of all hybrid rice production (nearly 17 Mha) originated from a single source of cytoplasmic male sterility, i.e. the wild abortive (WA) source. The genetic uniformity of modern rice varieties and hybrids results in the crop being highly vulnerable to outbreaks of numerous pests and diseases.

There is also concern that the spread of high-yielding varieties with limited genetic diversity erodes the diversity of rice. Few traditional varieties are cultivated at present and many may be lost as farmers adopt improved material. There is also evidence that outcrossing of improved varieties erodes genetic diversity. The loss of genetic diversity is a consequence of advances in agriculture, but it can also be argued that the contribution to world food security made by high-yielding varieties far surpasses the potential value of relying on traditional germplasm.

It is surprising that a breeding programme operating out of the Philippines can identify genotypes superior to those found in many national programmes working within the local production ecology. One interpretation is that many national breeding programmes have not yet reached the stage where they can effectively utilize introduced breeding lines to cross into local material, but instead they are dependent on introduced material for new varieties. The end result of this deficiency is the widespread adoption of elite breeding material from a single source and subsequent genetic uniformity. The apparent weakness of the national breeding programme has serious implications for the promotion of advanced technologies, such as hybrid rice and transgenic material. The problem of genetic uniformity is not limited to developing countries. United States rice is notorious for its narrow genetic base, and nearly all production in Australia utilizes the same variety that is widely used in California. Rice production in Latin America is also built upon a narrow genetic base.

Table 2. Nutritional aspects of rice consumption in selected countries

Country	*Consumption*	*Energy*	*Protein*	*Fat*	*Calcium*	*Iron*	*Niacin*	*Zinc*
	(g/day)	*(% of recommended daily intake)*						
Bangladesh	441	76	66	18	3	8	25	30
Brazil	108	14	10	0.8	<1	2	6	8
China	251	30	20	17	2	4	14	17
India	208	31	24	4	1	4	12	15
Indonesia	414	51	43	8	3	7	24	29
Myanmar	578	74	68	20	4	10	32	40
Philippines	267	41	30	5	2	5	14	17
Sri Lanka	255	38	37	3	2	5	14	17
Thailand	285	43	33	5	2	5	17	21
Viet Nam	465	67	58	14	3	8	27	34

Nutritional Quality

The relatively low nutritional quality of rice is a concern, especially in areas where rice consumption is very high. Rice is the main source of energy and is an important source of protein providing substantial amounts of the recommended nutrient intake of zinc and niacin (Table 2). However, rice is very low in calcium, iron, thiamine and riboflavin and nearly devoid of beta-carotene. There appears to be some genetic variation for iron and zinc content in rice, which may offer an opportunity for improving the nutritional value of rice as far as these minerals are concerned. "Golden Rice" (vitamin A-enriched rice) may contribute to alleviating vitamin A deficiency in the future.

Limited Resources

There is limited potential for expanding the production area due to increasing competition for land and water from industrialization and urbanization. Irrigated rice is cultivated annually on approximately 75 Mha, i.e. about half of the total area planted to rice. However, due to its high rate of productivity, irrigated rice is responsible for approximately 75 percent of total rice production. The problem with irrigated rice is its abundant consumption of water, which surpasses the actual needs of the crop. In Asia, irrigated rice consumes 150 billion m^3 of water, resulting in water-use efficiency of approximately 20 000 m^3/ha. Assuming an average yield of 5 t/ha, the water productivity of irrigated rice is only 0.15 kg of milled rice per m^3 of water. The current price of milled rice on the international market is approximately US$250 per tonne, which means water productivity of just US$0.038 per m^3. The low productivity of irrigated rice makes the crop noncompetitive compared to other uses of water.

It has been estimated that a decrease of 10 percent in water use for irrigated rice would save approximately 150 000 million m^3, which is the equivalent of approximately one-quarter of all the fresh water used worldwide for non-agricultural activities. Numerous studies have shown that irrigated rice can be easily cultivated using between 8 000 and 10 000 m^3/ha (i.e. approximately 50% of current use) without affecting yield. The major difficulty with conserving water is that the water is not priced properly. Where a charge is actually imposed, most irrigation schemes charge the user by area and not by volume of water consumed. Such a system provides no economic incentive to conserve water.

There are other proposals for reducing irrigated rice water consumption, including: limiting rice cultivation to the rainy season only; adoption of more water-efficient varieties (C4-type plants); promotion of upland rice; and use of biotechnology tools to induce drought tolerance in rice. However, all of these alternatives have a cost. Irrigated rice is most productive during the dry season, little genetic variation exists in rice in terms of water use (rice is a semi-aquatic plant) and upland rice provides low and unstable yields. The introduction of C4-type photosynthesis addresses only a small portion of the overall water use in irrigated rice production.

However, much can be done through simple conservation: maintaining supersaturated soil conditions only during cultivation of the crop; eliminating or severely reducing land preparation in water (puddling); and keeping water in the field by reducing discharges. In addition, more efficient delivery systems that reduce conveyance losses are required. Good weed management must be part of this strategy. Improved efficiencies in rice production are urgently required in order to compete with alternative uses of these resources.

Environmental Concerns

Agricultural activity is the main user and abuser of natural resources and there is worldwide pressure for more benign agricultural activities. Future production will be subject to public scrutiny and under pressure to comply with numerous international agreements, in particular the World Trade Organization's Agreement on Agriculture and the UN Framework Convention on Climate Change. Irrigated rice production is particularly vulnerable to emerging environmental regulations, due to the excessive use of irrigation water, indiscriminate use of pesticides and inefficient use of fertilizers. Rice is a prime suspect for the high level of carbon dioxide, methane, nitrous oxide and ammonia emissions. Carbon dioxide is emitted during the burning of crop residue - a standard practice in much of the world. Environmental regulations in California limit the burning of rice straw to no more than 25 percent of the cultivated area per year. Most countries, however, have not developed regulations regarding the burning of crop residue, although they may be compelled to do so by international trade agreements.

Debate in the international community has recently shifted the emphasis from carbon dioxide to more potent gases. Methane emission is peculiar to irrigated rice due to the long periods of flooding and the anaerobic decomposition of incorporated organic matter. It is estimated that irrigated rice accounts for 20 percent of the global emission of methane. Methane is approximately 20 times more potent than carbon dioxide as a greenhouse gas. Approximately 90 percent of methane emissions are emitted via the plant and there are reports of genetic differences in rice in terms of methane emissions. Although there is little information from field studies concerning cultural practices and methane emissions from rice, one possibility is the reduction of methane by short dry fallow or by rotation with an upland crop to allow organic matter to decompose under aerobic conditions before subjecting the soil to anaerobic conditions for irrigated rice cultivation.

Nitrous oxide is derived mainly from mineral nitrogen fertilizer. In irrigated rice, large quantities of nitrous oxide are emitted when soils are left to dry following application of urea and flooding. Continuous water maintenance after applications of urea fertilizer on dry soil can significantly reduce nitrous oxide emissions. Ammonia is also derived from mineral nitrogen fertilizer. Approximately one-sixth of global emissions of ammonia is attributed to fertilizer use in crop production. In irrigated rice, ammonia emission is mainly the result of inefficient use of urea. The principal problem is the application of urea in water or on mud, which results in approximately 70 percent of the urea volatilizing as ammonia. This rate can be reduced by at least 30 to 50 percent without affecting yield by adopting efficient but simple methods of N fertilizer management.

Rice Policy and International Obligations

Historically, most governments in the major rice-producing and rice-consuming countries developed policies to maintain paddy prices stable for consumers in urban centres and provide subsidies to rice farmers. Such policies led to a continuous expansion of rice production and held market prices within the purchasing power of low-income consumers. In major rice-exporting countries, policies favoured the export market, often at the expense of the producers. Current world prices are the lowest for several decades. Exports from Thailand, the world's biggest exporter, reached record levels in 2001/02, despite the fact that premium quality white rices were being sold for only US$178/tonne. This is made possible by the strong government support in Thailand. Global depression in rice prices is also a concern in Latin America where several governments have instituted trade barriers to protect national production. In the United States and Western Europe, rice is one of the most highly subsidized crops with massive government expenditure to support rice growers. In addition to such government support being expensive, policies to support the rice sector are limited by commitments under the Uruguay Round on Agriculture and structural adjustment programmes. Consequently, sustainable rice production in the long term would require improvement in production efficiency.

Technology Transfer

Finally, there are a plethora of institutional and policy factors requiring significant change if future requirements are to be fulfilled. There will be an increased need for technical assistance in order for farmers to adopt more efficient practices. Where public sector-supported technology transfer programmes and other support services suffer from institutional inertia and are inefficient and ineffective, new paradigms are required to provide and finance support services, assist in technology transfer and improve processing and marketing.

Addressing Future Trends

Scientific development and new issues in the rice industry sector, on the other hand, offer opportunities which, if properly harnessed, can lead to sustainable production. Urbanization offers new approaches to rice marketing, and with rising incomes there is an increased demand for quality rice, thus providing opportunities for market-driven rice-based production systems. Biotechnological tools and other advanced breeding methods could help to generate rice varieties with higher-yielding potential and which are more tolerant or resistant to biological and abiological stresses, as well as being more nutritive. Holistic and integrated crop management systems increase yield and reduce costs and environmental degradation by increasing the efficiency of rice-growing activities. Higher income from rice production may be achieved by reducing losses during harvest and post-harvest operations as well as via the diversification of the intensive rice-rice system.

Development of the New Generation of Rice Varieties

Increasing the genetic yield potential of rice has been proposed by IRRI (International Rice Research Institute) as a means for stimulating yields with subsequent research into "New Plant Types" (NPT). This concept was first developed in the early 1990s based on crop modelling with the stated goal of identifying NPTs with a yield potential of between 12 and 15 t/ha by the year 2002. However, progress with NPTs has been slow.

Hybrid rice is the most significant technology to emerge since the identification of dwarf plant types. Hybrids have consistently shown a 15 to 20 percent increase in yield compared to conventionally bred varieties. Scientists in China initiated hybrid rice breeding efforts in the late 1960s and the first commercial hybrid was introduced in 1974. In 2000, it was estimated that hybrid rice was cultivated on approximately 16 Mha in China, 300 000 ha in Viet Nam, 150 000 ha in India and 30 000 ha in Bangladesh. There are also minor plantings of hybrids in the Democratic People's Republic of Korea, Myanmar and the Philippines. A commercial company in the United States has also been active in hybrid research, with the objective of identifying hybrids suitable for production in the temperate regions of the Americas, i.e. the southern United States, southern Brazil, Uruguay and Argentina. Currently, almost all production is based upon three-line hybrid varieties. China has recently been promoting two-line hybrids, and in 2001 these hybrids covered approximately 2 Mha.

Table 3. Estimates of the yield gap in irrigated rice in selected countries

Country	*Improved yield (t/ha)*	*Farmers' yield (t/ha)*	*Yield gap (t/ha)*	*Yield gap (%)*
India	5.8	2.8	3.0	52
Philippines				
wet season	5.7	3.7	2.0	35
dry season	7.8	3.9	3.9	50
Latin America	6.3	4.6	1.7	27
Mauritania	7.2	4.1	3.1	43
Senegal	6.8	5.1	1.7	25
Mali	8.2	6.1	3.1	26
Burkina Faso	7.0	4.6	2.4	34
IRRI (9 countries)	8.2	3.7	4.5	55
Mean	7.0	4.3	2.7	46

Source: FAO, 2001.

Despite the progress made, several factors have tempered the widespread adoption of hybrids, and the major constraint is seed production. F1 seed production in China was reported to be only 0.41 t/ha in 1976, but technological advances increased seed yield to 2.5 t/ha in 1995. Chinese scientists reported higher F1 seed yields (2.5-3 t/ha) from the two-line system. However, outside China, F1 seed yields remain low. In both Viet Nam and India, F1 seed yields were reported to be as low as 1.3 to 1.7 t/ha), resulting in higher prices of hybrid seed and consequently high costs of commercial hybrid rice cultivation. Hybrid adoption is, therefore, still limited to areas where labour wages are low and transplanting is practised. The persistent low yield of hybrid seed production outside China - despite considerable investment and efforts by FAO, UNDP (United Nations Development Programme), IRRI, ADB (Asian Development Bank) and National Research Systems during the 1990s - suggests that the wide adoption of hybrid rice outside China still needs the identification of, or the breeding for, CMS lines with a good ability to maintain pollen sterility and a high outcrossing rate.

In summary, efforts to increase yield by raising the yield barrier with NPTs or hybrid rice are still hindered by inherent technological difficulties. Biotechnological tools and methodologies may provide a solution to these difficulties. The most important achievement in rice biotechnology is the sequencing and mapping of the rice genome, carried out by the International Rice Genome Sequencing Project (IRGSP) and other public and private institutions. Knowledge of the rice genome could have an important role in the breeding of new rice varieties, including the transfer of genes from other crops and organisms to rice.

At present, there is only one reported variety of rice in commercial production developed using transgenic procedures. However, there are several reports of material in the testing stage. Most involve herbicide resistance or use of the *Bt* gene; herbicide-resistant mutants will soon be released. The use of genetically modified organisms (GMOs) is still highly controversial; nevertheless, new traits of importance to both consumers and producers are likely to be available in the forthcoming decade. Rice-producing countries will have to decide on how and to what degree these innovations will be adopted. Biosafety analysis is important, but this is not only a GMO-related consideration.

Narrowing the Yield Gap

Most existing high-yielding varieties have a genetic yield potential of approximately 10 t/ha. Under excellent management in farmers' fields, these varieties frequently produce 7 to 8 t/ha, but the average yield by producers is about half this amount. The last session of the IRC recommended that FAO undertake a careful analysis of this yield gap. In September 2000, FAO organized the Expert Consultation on Yield Gap and Productivity Decline in Rice Production. Numerous results from on-farm field studies showed that the yield gap in the developing world was approximately 46 percent - the equivalent of approximately 2.7 t/ha. The yield gap is most apparent in irrigated ecologies for various reasons: the production environment is more suitable for high yields; most growers are already using improved genotypes; and there is a lot of technology available for irrigated rice production. An increase in production of approximately 130 Mt of milled rice could be achieved by bridging the yield gap in the irrigated rice areas of Asia (which cover 72 Mha). Narrowing the yield gap in irrigated rice, therefore, merits major attention.

The gap results from numerous deficiencies arising mainly from inadequate crop management practices. While improved crop management technologies are available, few have been introduced, tested and modified to suit local conditions. Deficiencies in crop management are often the result of inadequate technology transfer. Most technology transfer efforts are provided by public sector entities characterized by limited funding, inadequately trained staff and little incentive for improvement. Furthermore, many extension agencies suffer from institutional inertia. New approaches to technology transfer are essential.

There are examples of innovative means of providing services to growers without massive public sector support. Rice producer associations in Latin America have organized themselves into FLAR (Latin American Fund for Irrigated Rice) with established collaborative programmes that are international in scope and deal with common problems - thereby avoiding duplication of efforts and facilitating information exchange. In Australia, the "*Ricecheck*" technology transfer programme has been highly successful in increasing national yields by bridging the yield gap.

There are important differences between "knowledge-based" and "seed-based" technologies. Improved features incorporated into the genes of an elite cultivar are relatively fixed and visually evident and farmers can easily distinguish between disease-resistant and susceptible varieties and other improved phenotypic traits. Additionally, once growers accept seed-based technologies, they can continue using them without further assistance or additional costs (unless they are based on hybrids). On the other hand, it is often difficult for growers to appreciate the benefits of improved water management, which significantly reduces water usage. Similarly, farmers rarely understand how important it is to maintain biological equilibrium in pest management. Farmers can observe the response to N fertilizer but not large losses that occur via volatilization due to inappropriate applications.

Rice Integrated Crop Management to Narrow the Yield Gap

Limitations in crop management do not exist in isolation but are interlinked. For example, increased seedling vigour resulting from the use of high quality seeds will not affect yield if the crop is inadequately fertilized. Similarly, the crop cannot respond to improved fertility if weeds are permitted to compete. It is also very difficult to obtain high fertilizer efficiency without proper water management. These examples illustrate the importance of developing an integrated approach. However, many of these limitations are often addressed as single issues, resulting in programmes such as Integrated Nutrient Management, Integrated Pest Management and Integrated Water Management. When crop management constraints are integrated into a unified technology transfer programme rapid progress can be made. The integrated *Ricecheck* programme in Australia addresses ten key management factors in a holistic fashion. This integrated programme was instrumental in increasing the country's average yield from 6.5 t/ha during the 1980s to 8.9 t/ha in the late 1990s. During this period the yield gap was reduced from 45 to 33 percent. This experience is an example of the benefits that can be derived from an integrated approach and an innovative extension service.

The *Ricecheck* programme is a decision-making process in which a series of limitations to production have been identified and integrated into a production programme. For Australian conditions, critical management practices include crop establishment, planting times, N management, weed control and water management. Farmers are encouraged to follow best management practices based on "critical checks" known to have a major impact on yield. In addition to the development of the "critical checks" concept, the approach to technology transfer is unique and based upon establishing farmer groups that not only accelerate technology transfer but also serve as a feedback mechanism for researchers and extension workers.

Improved N management is another good example of the need for an integrated approach to crop management. In the transplanting system - such as that practised in most of Asia - N management is particularly difficult, because the land is prepared in water, the crop is seeded in mud and humid conditions prevail throughout crop development. The application of urea (the most common source of N fertilizer) in water or moist soil results in large losses, normally between 70 and 80 percent of the N applied. These losses produce low N efficiency (usually in the range of 5-15 kg of grain/kg of N) and contribute significantly to greenhouse gas emissions, such as nitrous oxide and ammonia.

Table 4. Mean N fertilizer rate and N fertilizer efficiency in five Asian countries

Site/country	*N fertilizer rate (kg N/ha)*	*N fertilizer efficiency (kg grain/kg of N)*
Mekong Delta, Viet Nam	46	8.6
Central Luzon, Philippines	136	12.5
Northwest Java, Indonesia	108	5.3
Central Flood Plains, Thailand	114	11.4
Tamil Nadu, India	115	15.3
Mean	112	10.6

Recent studies by FLAR in Columbia demonstrate the merits of incorporating N into dry soil prior to planting: N losses were only 20 to 30 percent with efficiency levels of 25 to 30 kg of grain per kg of N. These studies clearly illustrate that N efficiency can be increased threefold without additional costs, simply by timely applications under conditions that reduce losses (dry soil), followed by adequate water and weed management.

Reducing Post-harvest and Processing Losses

Higher yields will only lead to increased production if accompanied by improvements in post-harvest operations. Post-harvest handling of rice in much of Asia has not improved for decades. Hand-harvesting and threshing are common, rudimentary grain-drying prevails and rice is poorly stored. Results from surveys conducted by FAO show that losses from these operations amount to between 10 and 37 percent of the total harvest (Table 5). Technology is available for reducing these losses to more acceptable levels.

Losses from improper drying are high in much of Asia. Sun-drying is the most common method and it provides little control over the rate of drying. Much of the grain is either dried in the field in windrows or spread out on surfaces (roadsides) after threshing. Lack of control of moisture loss during drying results in fissured grain which breaks during milling. Additional losses in the milling process are caused by the inadequate technical performance of milling equipment. For example, the popular Engleberg-type single pass mill is notorious for breaking grain during milling, and milling yields are often only 50 percent, compared to the 70 percent milling yields achieved with modern milling equipment.

Table 5. Estimates of losses in Southeast Asia during various stages from harvest to processing

Process	*Range of losses (%)*
Harvest	1 - 3
Handling in the field	2 - 7
Threshing	2 - 6
Drying	1 - 5
Storage	2 - 6
Transport	2 - 10
Total	***10 - 37***

Normally, the price ratio of intact or three-quarter grain to broken grain is 3: 1. Consequently, high grain breakage means large financial losses. For decades, most Asian countries were concerned with rice self-sufficiency and limited attention was given to quality. However, in recent years consumer purchasing power has increased and quality rice is preferred. In addition, the export market demands high quality rice. Improving milling yields and reducing breakage are readily amendable factors, but major private sector investment is required.

Besides improving the drying process, there are also other technologies available for reducing milling losses. A new genetic trait labelled "tolerance to delayed harvest" permits grain to dry in the field to low levels (19%), thereby allowing flexibility in harvesting without inducing losses due to breakage during milling. Tolerance to delayed harvesting has been found in two commercial varieties in the United States and several sources have been identified in tropical and temperate regions of Latin America. Tolerance to delayed harvesting is becoming a major breeding objective in Latin America. This genetic trait has not been used in Asian breeding programmes but it could be a valuable trait.

Improving the Incomes and Livelihood of Rice Farmers

The intensive cultivation of irrigated rice has been advocated as a means of increasing production per unit area to meet food security. Many countries promoted early varieties in the 1980s to facilitate three rice crops per year. However, rice is a labour-intensive crop in much of Asia. Labour requirements in five major rice-producing countries range from 243 person-days/ha in Bangladesh, to 195 in India, 156 in Indonesia and between 60 and 80 in the more mechanized production systems of the Philippines and Thailand. Combining information on average yields in these areas with government support prices and labour requirements permits a gross estimate of the return-to-labour. The return-to-labour is estimated at: US$1.53/day in India, US$2.57/day in Bangladesh, US$6.06/day in Thailand, US$5.58/day in Indonesia and US$7.28/day in the Philippines (6able 8). Such low returns-to-labour are feasible only when no alternative employment is available.

Table 6. Estimates of gross returns on labour in irrigated rice in selected Asian countries

Country	*Manual labour (person-days/ha)*	*Price of rice (US$/t)*	*Average irrigated yield(t/ha)*	*Return onlabour (US$/day)*
Bangladesh	243	162	4.6	2.57
India	195	116	3.6	1.53
Indonesia	156	187	5.3	5.58
Philippines	82	211	3.4	7.28
Thailand	64	127	4.0	6.06

Return-to-labour is directly related to yield. Improvements in yield combined with greater production efficiency increases the return-to-labour. One of the primary benefits of narrowing the yield gap would be a greater return-to-labour. Recently, because of sufficient national rice production and with restrictions imposed on irrigation water, several countries are currently discouraging intensive rice cultivation. Also, markets for more highly lucrative crops, such as vegetables, offer the opportunity for farmers to increase their incomes. Crop diversification within the irrigated rice system is often advocated as a means for stabilizing rice yields, reducing methane emissions and increasing farmers' incomes. In many cases, diversification is hindered by the lack of suitable crops that can be grown on heavy, ill-drained ricelands, especially during the rainy season. While crop diversification requires more attention, efforts must focus on identifying alternative marketable crops suitable for heavy soils.

Several practices exist for reducing intensive labour use in rice, such as direct seeding, which avoids the laborious task of transplanting. Direct seeding is the standard means for crop establishment in the Americas, and the technology (know-how) for direct seeding is, therefore, readily available. Other technologies for reducing laborious practices are also available, but they must be adjusted to local conditions. Farm labour is a major source of employment in several Asian and African countries; any attempt to reduce labour must, therefore, be in balance with social concerns, especially in relation to rural employment.

Status of the World Rice Market

Rice is a major food staple and a mainstay for the rural population and their food security. It is mainly cultivated by small farmers in holdings of less than 1 ha. Rice is also a "wage" commodity for workers in the cash crop or non-agricultural sectors. This duality has given rise to conflicting policy objectives, with policy-makers intervening to save farmers when prices drop, or to defend consumer purchasing power when there are sudden price increases.

Rice is vital for the nutrition of much of the population in Asia, as well as in Latin America and the Caribbean and in Africa; it is central to the food security of over half the world population, not to mention to the culture of many communities. Rice is therefore considered a "strategic" commodity in many countries and is, consequently, subject to a wide range of government controls and interventions.

Patterns of Production and Trade

During the 1990s, global rice production expanded at a rate of 1.8 percent per year - marginally above the population growth rate. By the end of the decade, it reached 400 million tonnes (Mt) in milled equivalent (Figure 2). Developing countries account for 95 percent of the total, with China and India alone responsible for over half of the world output (Figure 3). Most of the increase in the 1990s was sustained through productivity gains rather than land expansion. In recent years the tendency for yield growth to slacken has been cause for concern. Furthermore, competition for basic resources (in particular, land and water) from other agricultural and non-agricultural sectors, as well as the negative environmental impacts associated with rice cultivation, are expected to pose a serious challenge to the future development of the sector.

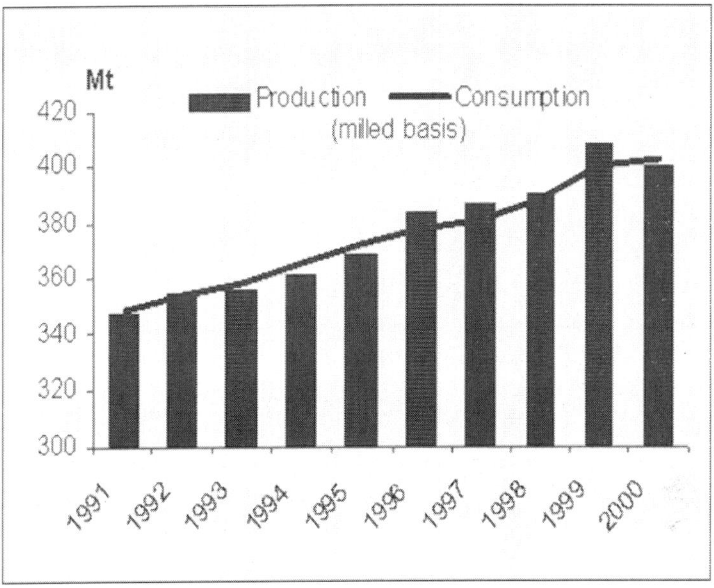

Figure 2. Global rice production and consumption

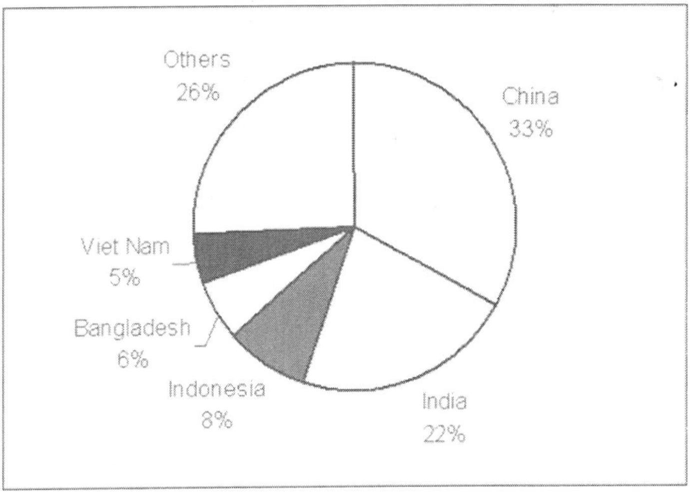

Figure 3. Major rice producers, 1998-2000

During the 1990s, global trade in rice expanded on average by 7 percent a year to about 25 Mt. Despite such dynamic growth, the international rice market remains thin, accounting for only 5 to 6 percent of global output. Unlike for other bulk commodities, the international rice market is segmented into a large number of varieties and qualities, which are not easily interchangeable because of strong consumer preferences. Ordinary *indica* rices are the most commercialized (some 80 percent of international trade by the end of the 1990s) followed by aromatic (Basmati and fragrant) rices at 10 percent, medium rices at 9 percent and glutinous rices at 1 percent.

Economic, Market and Institutional Integration

Developing countries are the main players in the world rice trade, accounting for 83 percent of exports and 85 percent of imports. The concentration is particularly high on the export side, since five countries (Thailand, Viet Nam, China, the United States and India) cover about three-quarters of world trade. This situation is in contrast to the fragmentation of import markets and the wide year-to-year fluctuations in individual countries' purchases, resulting from the fact that importers do not rely consistently on the international market for rice supplies, but only as a last resort to fill the gap caused by a production shortfall. Unlike for other important cereals, there is no major international future market for rice.

During the 1990s, Brazil, Indonesia, Saudi Arabia, Iraq and the Islamic Republic of Iran were the major destinations in the rice trade. In the last few years, however, following a shift in policies, imports by African countries have surged, providing a major stimulus to trade.

Because of rice's importance for food security and political stability, a significant proportion of trade is conducted by state trading enterprises, some of which are also obliged to procure or distribute rice domestically. This applies to both importing and exporting countries. However, in recent years, many of these state enterprises have lost their monopoly position and private traders have taken on greater responsibility for dealing with rice imports and exports.

Government-to-government transactions, which used to account for about half of world trade in the 1970s, are now estimated to represent less than 10 percent of the total. In the past few years, however, they have regained popularity, as low international prices have incited or compelled governments to play a more active role in trade, either to gain bargaining power or as an indirect means of sustaining producer prices. These transactions often take place under conditions that can barely be matched by private traders, especially as far as credit is concerned.

General Thrust of Rice Government Policies

Government support to producers in developing countries concentrates mainly on: research in improved or hybrid rice varieties, investments in irrigation, preferential credits, extension and distribution of improved seed. Intervention to influence prices is also common - through procurement purchases or releases from stocks, or through changes in trade policies. In developed countries, much assistance to the sector is conveyed through direct payments and through price support.

Government often plays an important role in the first phase of the marketing cycle, by procuring paddy at minimum producer prices.

In general, the involvement of the public sector in paddy processing and rice distribution is more limited. However, some governments oblige millers to purchase paddy at a predetermined price and to charge fixed mark-ups at each stage of the marketing process, while others distribute rice at fixed retail prices. It is common practice to manage rice stocks or to adopt trade policy measures in order to stabilize domestic market prices.

Trade measures, especially tariffs, are widely used to protect domestic rice markets. Despite the relatively high WTO (World Trade Organization)-bound tariffs, rice imports are often subject to "Special Safeguards" in country schedules. The role of state companies in managing international rice flows is also important, although such companies do not usually have monopoly privileges and they share their trade functions with the private sector. Many commercial transactions are conducted through government-to-government deals, the terms of which are not usually released to the market, contributing to poor transparency. Restrictions on exports of paddy or husked rice are very common, reflecting an endeavour to promote domestic rice processing.

Because of the importance of rice as a staple food, many governments maintain minimum food reserves to ensure food security. In addition, countries engaged in rice distribution schemes and producer price support usually keep large rice inventories in public storage facilities.

Since rice is one of the most protected traded commodities, there is considerable scope for further market liberalization. However, because of its importance in terms of food security, income generation and political stability, governments may be reluctant to loosen their control over the sector. Moreover, rice is central to the concepts of food security and multifunctionality - as promoted by a number of countries for consideration in the Doha new round of multilateral trade negotiations.

Current Issues and Problems

Falling international prices have been the principal cause for concern in the last few years, for both importing and exporting countries. The slide in world quotations was a reflection of the dynamic growth in global production since the mid-1990s, following the implementation of expansionary policies in a large number of countries. Although world paddy production has fallen in the past 2 years, supply releases from stocks have kept the downward pressure on prices.

Although genetically modified rice varieties have been developed (to enhance nutritious characteristics, e.g. "Golden Rice", or for adaptation to extreme growing conditions, e.g. varieties tolerant to salty water), the issue of their acceptability worldwide has not yet gained prominence, because rices produced from such varieties are not yet widely traded. More importantly, concerns have arisen regarding the use of the "Basmati" rice denomination and claims of biopiracy on fragrant rice genes.

Rice production sites are often the natural habitat for a wide variety of birds and plants. Water management in ricelands also ensures a soil desalination process essential to the maintenance of land fertility. As a result, environmental concerns frequently come to the fore in defence of the sector, especially in developed countries.

Review of Rice Policies from a WTO Perspective

Market Aaccess

Bound tariffs

Bound tariff rates for rice are generally high, often over 100 percent. Some countries (India, the Dominican Republic and Panama) recently renegotiated tariff bindings to higher than the originally agreed levels, alleging that the latter provided insufficient protection for their domestic sectors. Japan, the Republic of Korea, the Philippines and, more recently, Taiwan Province of China have invoked deferred tariffication under the "Special Treatment" provisions of Annex 5 of the Agreement on Agriculture. From the point of view of an international rice exporter, deferred tariffication is favourable, because it is associated with the opening of a larger preferential tariff quota (8 percent of base-period consumption at the end of the implementation period, instead of 5 percent for developed and 4 percent for developing countries).

Applied tariffs

Applied tariff rates are often set equal to bound tariffs when international prices are low. However, their level tends to change frequently, depending on the domestic and international market situations, contributing a high level of uncertainty to the world rice economy. Moreover, variable import duties (e.g. price band mechanisms) are popular in some countries in Central America and the Caribbean and in South America.

Tariff escalation

Tariff escalation is less important for rice than for other commodities, because of the relatively small volume of trade of paddy or unhusked rice. Nonetheless it should be noted that tariffs increase according to the processing stage and are much lower for paddy or husked rice than for milled rice.

Special Safeguards (SSG)

Several countries have reserved the right to use the Special Safeguard (SSG) provisions; however, only rarely have they been invoked for rice.

Tariff rate quotas

Eleven countries have made commitments to grant minimum access through preferential tariff rate quotas (TRQ) under the 1994 Uruguay Round Agreement (URA). Mainland China and Taiwan Province of China upon accession in 2002 also agreed to the opening of a preferential tariff rate quota. These quotas are sometimes subject to high tariffs (e.g. 80% in Colombia, 90% in Indonesia and 177% in Morocco) and preferential tariff quotas often constitute the only means for accessing important markets (EC, Japan, Republic of Korea, Taiwan Province of China etc.) where out-of-quota imports are subject to prohibitive tariff rates. In the case of mainland China, the initial import quota agreed is high (almost 4 Mt in 2002), with the in-quota tariff set at just 1 percent.

The problem of underfill of TRQs exists, especially for highly competitive, traditional exporting countries, such as Thailand. More recently, mainland China failed to import sizeable volumes of rice under its preferential access scheme - mainly a reflection of the low domestic rice prices compared to world quotations, partly induced by the release of large quantities of rice from government inventories. The administration of TRQs is often entrusted to state trading enterprises or other government agencies. There are a variety of methods for allocating the quotas to external suppliers: on the basis of tenders in Japan and the Republic of Korea; on a first-come, first-served basis in the Philippines; or on the basis of historical performance in the EC. The systems in place for allocating TRQs among different exporting countries do not appear to have caused substantial problems.

Non-tariff barriers

Phytosanitary barriers are less important for rice than for other commodities. Certain restrictions on rice imports from Viet Nam and Thailand are in force in several Latin American and Caribbean countries on phytosanitary grounds, influencing the pattern of trade flows into the region.

State trading is a characteristic feature of the world rice economy, since a large number of exporting and importing countries rely on government trade agencies to sell or purchase rice on the international market. These bodies may hold a trade monopoly on rice exports or imports (as is the case in the Philippines, Malaysia and China), or they may make rice transactions alongside private traders (Indonesia, Thailand, Viet Nam and Myanmar). Dealings through public agencies are not usually made openly and the quantities, prices and other conditions of the exchange are often kept secret. Lack of transparency is therefore often associated with state trading. However, lack of transparency also characterizes trading by producer associations; for example, the rice export association in Australia does not even reveal the destination of exports.

State trading or direct government-to-government transactions are most prevalent in Asia, particularly among Near East countries, but also in Bangladesh, China, Indonesia, the Philippines, Malaysia, Myanmar, Thailand and Viet Nam. It is not common in Latin America, although some governments do play an important role in rice trade using licensing and import permits rather than direct state trading.

Export Measures

Export subsidies

Export subsidy reduction commitments have been made under the Uruguay Round Agreement on Agriculture (URAA) by Colombia, Indonesia, Uruguay, the EC and the United States. The actual use of export subsidies has fallen short of the aggregate ceiling, although information is difficult to get even from the WTO. Proposals for further reduction commitments are likely to meet opposition from the EC. Other issues have arisen in relation to export competition in rice, in particular the granting of export credits by the United States of America. It should be noted, however, that export credits are also commonly used in government-to-government deals, although there is little information available in connection with such practices.

Export subsidies have been used by India since mid-2001 to promote exports of rice held by the government Food Corporation of India. According to the WTO, India is not eligible to use export subsidies on rice, but the country claims that under the URAA (Article 9-4) the country is exempt from commitments on export subsidies for marketing, processing and transportation. While this position is questionable, the country has to date not been challenged on that account by other countries in the WTO.

Export restrictions

Restrictions on the export of paddy and unhusked rice are also applied by a large number of exporting countries as a way of protecting the milling industry. This strategy reduces the flexibility of importing countries trying to promote value-adding processing industries by importing non-milled rice.

Domestic Support

Amber box domestic support

Developed countries completed their Aggregate Measurement of Support (AMS) reduction commitments in 2000, mainly through cuts in price support. Such cuts have been associated with a rise in compensatory payments to rice producers (classified under either the "blue" or the "green" box), which have been particularly important in the EC, Japan and the United States of America. The shift from price to income aids has not been accompanied by a major fall in production and some developed countries now find themselves with large rice stocks. Support to the rice sector accounts for a very high proportion of the total AMS in Japan and the Republic of Korea. These countries are expected to resist proposals for a reclassification of policies (e.g. from the green or blue box to the amber box). They are also likely to oppose further reductions in the AMS on the grounds of concern for national food security and preservation of the countryside, from an environmental, cultural and social perspective.

Few developing countries have submitted a base AMS, and few are therefore subject to reduction commitments. Most developing countries still have ample scope for increasing their assistance to the sector - should they choose to do so - under the "de minimis" provision. Only very sizeable reductions in the de minimis ceilings could negatively affect rice producers in those countries where the crop accounts for an important share of total agricultural outlays, i.e. many Asian countries and several Latin American and Caribbean countries. The proposal to raise the de minimis via special and differential treatment for least developed countries may have little effect, since 10 percent of the base production value already granted gives ample scope for domestic support to the commodity.

The impact of inflation and changes in exchange rates on current AMS may be of far greater importance for countries which submitted a base AMS in domestic currencies and where inflation is high.

Blue box domestic support

Decoupled, production-limiting payments are made to rice producers in the EC, Japan, the Republic of Korea and Mexico. Since 1999, they have been essential for allowing producers to weather the impact of low prices. They have been strongly criticized by other players in the rice market for not being truly "decoupled", and demands are made for their elimination or reduction by shifting decoupled income support and income safety nets from the blue or green boxes to the amber box, in order to make them subject to reduction commitments.

Green box domestic support

In the United States of America, considerable resources have been channelled to the sector through production flexibility contracts, retirement payments and payments for natural disasters, all of which are classified as green box measures. Since 2002, the new Farm Bill has endogenized the counter-cyclical and emergency payments which had been previously been provided through the 1996 US Farm Bill (FAIR ACT) on an ad hoc basis;

they are classified as green box measures. There is a prevailing tendency generally to promote non-commodity specific programmes, such as producer insurance schemes, also in developing countries.

A number of developing countries support the inclusion of a "Food Security" box which would permit the exemption of certain policies from reductions commitments. This would contain, inter alia, poverty alleviation measures and product-specific support for low-income farmers. It would be of very high relevance to rice production in a large number of countries, especially India.

Other Issues

The "multifunctionality" of agriculture in terms of environmental, social and cultural concerns is being used to defend the permanence of blue and green box payments. In Japan, most of the emphasis on multifunctionality and food security is in relation to rice. In developed countries where rice is a non-marginal crop, the elimination of blue or green box support would considerably impair the sector.

Rice production sites are often the natural habitat of a wide variety of birds and plants. Water management in ricelands ensures that the soil desalination process essential to the maintenance of land fertility takes place. Environmental concerns are consequently a frequently used weapon in defence of the sector.

Food safety is not particularly relevant to rice, although there is increasing concern regarding GMOs (genetically modified organisms). While some rice varieties are being developed with new genes (e.g. carotene-enriched rice), they are not yet traded internationally.

An emerging issue which is of importance to the WTO is that of intellectual property rights over particular varieties of rice, in particular in relation to the Basmati and fragrant rice varieties developed in the United States of America. "Biopiracy" of the genes is suspected and it is feared that the new strains could compete with traditional Basmati and fragrant rice exports from India, Pakistan and Thailand.

There have also been issues regarding the use of certain denominations, such as "Basmati" or "Jasmine". India and Pakistan are now trying to have the name associated with the geographical zone of production.

References

Agcaoili, M. & Rosegrant, M.W. (1992). *Regional trends and projections for food supply demand and trade in Asia*. Washington DC, IFPRI.

FAO. (2002). *World agriculture: towards 2015/30, an FAO study*. Rome.

Flinn, J.C. & De Datta, S.K. (1984). Trends in irrigated rice yields under intensive cropping at Philippine research stations. *Field Crops Res*., 9: 1-15.

Kenny, G. (2001). *Nutrient impact assessment of rice in major rice consuming countries*. FAO-ESNA Consultancy Report.

Pulver, E. & Nguyen, V.N. (1998). Sustainable rice production issues for the third millennium. In *Proceedings of the Nineteenth Session of the International Rice Commission*, Cairo, Egypt, p. 32-43.

Tuong, T.P. & Bouman, B.A.M. (2002). Rice production in water-scarce environments. In *Proceedings of International Conference on Irrigation Water Management*. USDA. 2002. ERS, Database.

Yuan, L.P. & Fu. (1995). *Technology of hybrid rice production*, p. 84. Rome, FAO.

8

Trends in World Wheat Market

Wheat is grown on more land area than any other commercial crop and is the most important staple food for humans. World trade in wheat is greater than for all other crops combined. Globally, wheat is the leading source of vegetable protein in human food, having a higher protein content than either maize (corn) or rice, the other major cereals. In terms of total production tonnages used for food, it is currently second to rice as the main human food crop and ahead of maize, after allowing for maize's more extensive use in animal feeds.

In 2010 world production of wheat was 651 million tons, making it the third most-produced cereal after maize (844 million tons) and rice (672 million tons). In 2009, world production of wheat was 682 million tons, making it the second most-produced cereal after maize (817 million tons), and with rice as close third (679 million tons).

Wheat was a key factor enabling the emergence of city-based societies at the start of civilization because it was one of the first crops that could be easily cultivated on a large scale, and had the additional advantage of yielding a harvest that provides long-term storage of food. Wheat contributed to the emergence of city-states in the Fertile Crescent, including the Babylonian and Assyrian empires. Wheat grain is a staple food used to make flour for leavened, flat and steamed breads, biscuits, cookies, cakes, breakfast cereal, pasta, noodles, couscous and for fermentation to make beer, other alcoholic beverages, or biofuel.

Harvested wheat grain that enters trade is classified according to grain properties for the purposes of the commodities market. Wheat buyers use these to decide which wheat to buy, as each class has special uses, and producers use them to decide which classes of wheat will be most profitable to cultivate.

Wheat is widely cultivated as a cash crop because it produces a good yield per unit area, grows well in a temperate climate even with a moderately short growing season, and yields a versatile, high-quality flour that is widely used in baking. Most breads are made with wheat flour, including many breads named for the other grains they contain like most rye and oat breads. The popularity of foods made from wheat flour creates a large demand for the grain, even in economies with significant food surpluses.

In recent years, low international wheat prices have often encouraged farmers in the USA to change to more profitable crops. In 1998, the price at harvest was \$2.68 per bushel. A USDA report revealed that in 1998, average operating costs were \$1.43 per bushel and total costs were \$3.97 per bushel. In that study, farm wheat yields averaged 41.7 bushels per acre (2.2435 metric ton / hectare), and typical total wheat production value was \$31,900 per farm, with total farm production value (including other crops) of \$173,681 per farm, plus \$17,402 in government payments. There were significant profitability differences between low- and high-cost farms, mainly due to crop yield differences, location, and farm size.

In 2007 there was a dramatic rise in the price of wheat due to freezes and flooding in the northern hemisphere and a drought in Australia. Wheat futures in September, 2007 for December and March delivery had risen above \$9.00 a bushel, prices never seen before. There were complaints in Italy about the high price of pasta. This followed a wider trend of escalating food prices around the globe, driven in part by climatic conditions such as drought in Australia, the diversion of arable land to other uses (such as producing government-subsidised bio-oil crops), and later by some food-producing nations placing bans or restrictions on exports in order to satisfy their own consumers.

Other drivers affecting wheat prices include the movement to bio fuels and rising incomes in developing countries, which is causing a shift in eating patterns from predominantly rice to more meat based diets (a rise in meat production equals a rise in grain consumption—seven kilograms of grain is required to produce one kilogram of beef).

In 2003, global per capita wheat consumption was 67 kg, with the highest per capita consumption (239 kg) found in Kyrgyzstan. In 1997, global wheat consumption was 101 kg per capita, with the highest consumption (623 kg per capita) in Denmark, but most of this (81%) was for animal feed. Wheat is the primary food staple in North Africa and the Middle East, and is growing in popularity in Asia. Unlike rice, wheat production is more widespread globally though China's share is almost one-sixth of the world.

In the 20th century, global wheat output expanded by about 5-fold, but until about 1955 most of this reflected increases in wheat crop area, with lesser (about 20%) increases in crop yields per unit area. After 1955 however, there was a dramatic ten-fold increase in the rate of wheat yield improvement per year, and this became the major factor allowing global wheat production to increase. Thus technological innovation and scientific crop management with synthetic nitrogen fertilizer, irrigation and wheat breeding were the main drivers of wheat output growth in the second half of the century. There were some significant decreases in wheat crop area, for instance in North America.

Better seed storage and germination ability (and hence a smaller requirement to retain harvested crop for next year's seed) is another 20th century technological innovation. In Medieval England, farmers saved one-quarter of their wheat harvest as seed for the next crop, leaving only three-quarters for food and feed consumption. By 1999, the global average seed use of wheat was about 6% of output.

Several factors are currently slowing the rate of global expansion of wheat production: population growth rates are falling while wheat yields continue to rise, and the better economic profitability of other crops such as soybeans and maize, linked with investment in modern genetic technologies, has promoted shifts to other crops.

World Wheat Harvested Area

The world wheat market is "segmented" into groups of countries with diverse productive capacities and levels of use, differing competitive positions in export import markets, and dissimilar prospects for the future as World wheat market participants. Given the critically important role of wheat as a necessity in international diets and foodeconomic systems, World wheat demand is projected to continue to grow along with World population.

Population growth is projected to grow from 6.852 billion in early March 2011, to 7 billion in 2012, and to 8 billion in 2027.

The harvested acreage of wheat in the World in the 2010/11 marketing year is estimated to be 222,220,000 hectares. This compares to an average World harvested wheat acreage estimate of 218,320,000 hectares since the 1987/88 marketing year, with average declines of 298,000 hectares per year over the most recent 24 marketing years.

The 10 largest wheat producing countries or regions provided an average of 79.9% of World wheat harvested acreage over the 1987/88 to 2010/11 marketing years. Over the most recent 3 marketing years (2008/09 to 2010/11), the 5 largest countries in terms of average World wheat harvested acreage were India (28.25 million hectares or mh), the European Union (25.92 mh), Russia (27.35 mh), China (24.07 mh) and the United States (20.67 mh or 51.08 million acres).

Australia and Canada have the largest historic variability in wheat area harvested relative to their average harvested area since 1987/88 measured in terms of the "coefficient of variation" or "cv", at 0.018 and 0.16, respectively. A coefficient of variation of 1.00 indicates that a measure of the variability of a variable (defined as the standard deviation) is cqual to the average value of the variable. When the cv value is less than 1, it meals that the standard deviation is less than the average of the variable.

In this case, Kazahkstan, China and the United States have lower levels of relative variation in harvested area of wheat over time than Australia and Canada, followed by Russia, India, Pakistan and Turkey, and the European Union. In comparison, aggregate World wheat harvested acreage has a relatively low level of relative variability since the 1987/88 marketing year.

From a market analysis perspective, these results indicate that the largest wheat acreage shifts over time have occurred in Australia and Canada, and that aggregate relative World variability in harvested wheat acreage is less than in most of the major wheat producing countries.

World Wheat Yields

Average World wheat yields in the 2010/11 marketing year are estimated to be 2.90 metric tons per hectare (mt/ha) (equivalent to 43.1 bushels per acre). This compares to an average World wheat yield estimate of 2.67 mt/ha (39.7 bu/ac) since the 1987/88 marketing year, with an average annual increase of 0.03 mt/ha (0.45 bu/ac) over the 24 year period.

Among thc 10 largest wheat producing countries or regions, the highest average wheat yields over the most recent 10 year period (2001/02 to 2010/11 marketing years) were produced in the European Union (5.14 mt/ha or 76.4 bu/ac), with yields trending upwards at 0.06 mt/ha (0.89 bu/ac) per year. The second highest yields over the most recent 10 year period were produced in China at 4.35 mt/ha or 64.7 bu/ac, with yields trending upwards at 0.12 mt/ha (1.8 bu/ac) per year. United States' wheat yields averaged 2.82 mt/ha (41.93 bu/ac) over the last 10 years, increasing at a rate of 0.05 mt/ha (0.74 bu/ac) per year.

Among the top 10 yielding countries since 1987/88, the Ukraine and Russia have the largest historic relative variability in wheat yields since 1987/88, with cv measures of 0.21 and 0.19, respectively. Argentina, and China, Pakistan and Canada have lower levels of relative variation in wheat yields over time, followed by Turkey, the United States, India and the European Union. Aggregate World wheat yields have exhibited a relatively low level of relative variability since 1987/88. This indicates that 2 of the major wheat producing countries in the Black Sea region (Ukraine and Russia) have the largest amounts of year to year uncertainty in regards to wheat yields, and that the varying yields in individual countries and regions tend to offset each other over time relative to aggregate average World wheat yields.

World Wheat Production

FAO's latest forecast for global wheat production in 2011 stands at 691 million tonnes, 6 percent above last year's level and some 6 million tonnes above the previous high, which was set in 2009. Even with some important southern hemisphere crops still to be gathered, the forecast for the 2011 world wheat crop is now quite firm and indicates a much larger harvest than expected at the start of the season. Although, as the season progresscd, crop prospects deteriorated in some major producing countries such as the United States and some EU countries, the predicted sharp recovery in production in the CIS countries from the drought-reduced level of 2010 has turned out even larger than expected, more than offsetting the downward revisions elsewhere.

In the EU, the aggregate wheat output is now estimated just over 1 percent up from 2010 at 138.6 million tonnes. Early season hopes for a larger crop were dashed when drought struck major producing areas from the

United Kingdom through France and Germany and into Poland. However, particularly favourable conditions led to unexpectedly good outputs in some eastern EU countries, particularly Hungary and Romania, that partially offset the drought-reductions elsewhere. In the rest of Europe, production in the CIS countries has exceeded earlier expectations, recovering sharply from the drought-reduced level of 2010 in the Russian Federation and Ukraine.

In North America, the United States recently completed its delayed spring 2011 wheat harvest, which was down by some 9 percent compared with 2010. In Canada, good summer weather improved the outlook for 2011 grain crops after an uncertain start of the season because of a late damp spring. Latest official forecasts now put the 2011 wheat harvest at just over 24 million tonnes, nearly 4.3 percent above last year's level.

Table 1. Wheat production: leading producers

	2010 estim.	***2011 f'cast***	***Change: 2011 over 2010***
	million tonnes		%
European Union	136.9	138.6	1.2
China (Mainland)	115.2	116.8	1.4
India	80.8	84.3	4.3
United States	60.1	54.7	-9.1
Russian Federation	41.5	57.0	37.3
Australia	26.3	26.2	-0.4
Canada	23.2	24.2	4.3
Pakistan	23.3	24.2	3.9
Turkey	19.7	21.8	10.7
Ukraine	17.0	22.5	32.4
Kazakhstan	9.6	22.2	131.3
Iran Islamic Rep. of	13.5	13.5	0.0
Argentina	14.7	13.0	-11.6
Egypt	7.2	8.4	16.7
Uzbekistan	6.7	6.3	-6.0
Other countries	56.1	57.4	2.2
World	651.8	691.0	6.0

Noite: Countries listed according to their position in global production (average 2009-2011).

In Asia, after concern about exceptionally dry conditions in some parts of China early in the season, the 2011 wheat harvest in the country has set a new record, up 1.4 percent from the previous record, which was set last year. Production in India and Pakistan has also risen to new highs. In the CIS group in Asia, production in Kazakhstan has recovered sharply from last year's drought-reduced harvest.

Elsewhere in the northern hemisphere, aggregate output in North Africa rebounded significantly from last year's drought-reduced level following upturns in the main producing countries. In the Near East, Turkey has harvested another bumper crop but, in most other countries, outputs were below average, reflecting late and erratic rainfall.

In the southern hemisphere, the bulk of the 2011 wheat crops are to be harvested between now and the end of the year. In South America, prospects in Argentina have deteriorated somewhat in recent weeks due to adversely dry conditions. Yields are expected to be well down from last year's records and given a similar area planted, the country's wheat crop is forecast to drop by almost 12 percent. In Oceania, prospects for the wheat crop in some eastern parts of Australia worsened during the latter part of the season, but conditions in Western Australia still favour a sharp recovery after last year's drought in that region. The country's aggregate wheat output is forecast to remain close to the overall good 2010 level.

In many parts of the northern hemisphere, the winter wheat crops for harvest in 2012 are already being planted, or are due to be sown in the next few weeks. Planting conditions are reported as generally favourable in most of the concerned areas, with the exception of the United States, where prolonged dryness in southern parts is hampering fieldwork, and Ukraine, where conditions are also adversely dry. With current wheat prices similar to their levels a year ago and utilization expected to outstrip supply in 2011/12, the crop should remain an attractive option for producers. As a result, farmers are expected to maintain, or even increase, the area under wheat. In the United States, early indications point to a considerable increase in wheat plantings for harvest in 2012, contrasting with the relatively small coverage in the past two years.

In Europe, plantings may also increase in the CIS countries, where farmers will be keen to continue benefiting from attractive prices and strong demand in the region after the huge production shortfall in 2010. In the EU, however, with other crops competing strongly for land, the wheat area is expected to remain relatively unchanged. In Asia, planting of winter wheat for harvest in 2012 is already underway or due to start in October in the main producing countries. Prospects in India are favourable but persistent dryness in parts of China and severe floods in the Sindh province of Pakistan could impact sowing in the affected regions.

Geographical Variation

There are substantial differences in wheat farming, trading, policy, sector growth, and wheat uses in different regions of the world. In the EU and Canada for instance, there is significant addition of wheat to animal feeds, but less so in the USA. The biggest wheat producer in 2010 was EU-27, followed by China, India, USA and Russian Federation.

The largest exporters of wheat in 2009 were, in order of exported quantities: United States, EU-27, Canada, Russian Federation, Australia, Ukraine and Kazakhstan. Upon the results of 2011, Ukraine became world's sixth wheat exporter as well. The largest importers of wheat in 2009 were, in order of imported quantities: Egypt, EU-27, Brazil, Indonesia, Algeria and Japan. EU-27 was on both export and import list, because EU countries such as Italy and Spain imported wheat, while other EU-27 countries exported their harvest. The Black Sea region – which includes Kazakhstan, the Russian Federation and Ukraine – is amongst the most promising area for grain exporters; it possess significant production potential in terms of both wheat yield and area increases. Black sea region is also located close to the traditional grain importers in the Middle East, North Africa and Central Asia.

In the rapidly developing countries of Asia, westernization of diets associated with increasing prosperity is leading to growth in per capita demand for wheat at the expense of the other food staples. In the past, there has been significant governmental intervention in wheat markets, such as price supports in the USA and farm payments in the EU. In the EU these subsidies have encouraged heavy use of fertilizer inputs with resulting high crop yields. In Australia and Argentina direct government subsidies are much lower.

In the Punjab region of India and Pakistan, as well as North China, irrigation has been a major contributor to increased grain output. More widely over the last 40 years, a massive increase in fertilizer use together with the increased availability of semi-dwarf varieties in developing countries, has greatly increased yields per hectare. In developing countries, use of (mainly nitrogenous) fertilizer increased 25-fold in this period. However, farming systems rely on much more than fertilizer and breeding to improve productivity. A good illustration of this is Australian wheat growing in the southern winter cropping zone, where, despite low rainfall (300 mm), wheat cropping is successful even with relatively little use of nitrogenous fertilizer. This is achieved by 'rotation cropping' (traditionally called the ley system) with leguminous pastures and, in the last decade, including a canola crop in the rotations has boosted wheat yields by a further 25%. In these low rainfall areas, better use of available soil-water (and better control of soil erosion) is achieved by retaining the stubble after harvesting and by minimizing tillage.

In 2009, the most productive farms for wheat were in France producing 7.45 metric tonnes per hectare. The five largest producers of wheat in 2009 were China (115 million metric tonnes), India (81 MMT), Russian Federation (62 MMT), United States (60 MMT) and France (38 MMT). The wheat farm productivity in India and Russia were about 35% of the wheat farm productivity in France. China's farm productivity for wheat, in 2009, was about double that of Russia. If India and Russia could adopt the farming knowledge and technology of France, the world production of wheat would be 40% higher with a farming area same as the area farmed for wheat in 2009.

In addition to gap in farming system technology and knowledge, some large wheat grain producing countries have significant losses after harvest at the farm and because of poor roads, inadequate storage technologies, inefficient supply chains and farmer's inability to bring the produce into retail markets dominated by small shopkeepers. Various studies in India, for example, have concluded that about 10% of total wheat production is lost at farm level, another 10% is lost because of poor storage and road networks, and additional amounts lost at retail level. One study claims that if these post-harvest wheat grain losses could be eliminated with better infrastructure and retail network, in India alone enough food would be saved every year to feed 70 to 100 million people over a year.

World Wheat Food, Seed & Industrial Use

World wheat food, seed and industrial (FSI) use in the 2010/11 marketing year is estimated to be 539,572,000 metric tons. This compares to average

World wheat FSI use of 504,751,000 metric tons since the 1987/88 marketing year, with average increases of 5,271,000 metric tons per year over the 24 year period.

The 10 largest wheat food, seed and industrial using countries or regions used an average of 71.5% of World wheat over the 1987/88 to 2010/11 time period. Over the most recent 3 year period (2008/09 to 2010/11 marketing years), the 5 largest countries or regions in terms of average World wheat FSI use were China (97.1 mmt), India (77.1 mmt), the European Union (67.8 mmt), the United States (27.2 mmt or 999 million bushels) and Russia (22.9 mmt).

Egypt and India have the largest historic variability in wheat FSI use relative to their average quantity produced since 1987/88, with cv measures of 0.15 and 0.13, respectively. Iran and Pakistan, and Brazil have lower levels of relative variation in wheat FSI use over time, followed by the United States, Turkey, the European Union, Russia and China. Aggregate World wheat FSI use has exhibited a relatively low level of production variability since MY 1987/88. Overall, annual wheat FSI use has been more variable in Egypt, India, Iran, Pakistan and Brazil than in other major wheat FSI use countries, and variation of aggregate total World wheat FSI use has been less than that among individual countries and regions since MY 1987/88.

World Wheat Feed Use

World wheat feed use in the 2010/11 marketing year is estimated to be 123,102,000 metric tons. This compares to average World wheat feed use of 109,464,000 metric tons since the 1987/88 marketing year, with average increases of 128 metric tons per year over the 24 year period.

The 10 largest wheat feeding countries or regions used an average of 91.9% of World wheat fed over the 1987/88 to 2010/11 time period. Over the most recent 3 year period (2008/09 to 2010/11 marketing years), the 5 largest countries or regions in terms of average World wheat feed use were the European Union (56.7 mmt), Russia (19.9 mmt), China (10.0 mmt), the United States (5.2 mmt or 192 million bushels) and Australia (4.3 mmt).

The Ukraine has an extremely large amount of relative annual variability in wheat feeding since MY 1987/88, with a cv measure of 1.79, followed by South Korea, the United States, Kazakhstan, Russia and Egypt.

China, Australia and the European Union have lesser amounts of relative annual variation in wheat feeding since MY 1987/88. Overall, annual wheat feed use has been extremely variable in a number of countries, likely owing to the relatively small proportion that feed use makes up of overall wheat usage, and how the coefficient of variation measure is affected by large proportional measures of variation relative to small average wheat feeding quantities.

PRICES

Lower Prices amid Higher Production

Larger than anticipated world production and a strengthening US Dollar contributed to the decline in international wheat prices in recent weeks. A recovery in the CIS countries and bigger crops than expected in northern Europe helped improve the supply outlook and put more downward pressure on prices. While at the beginning of the 2011/12 marketing season in June, the benchmark United States No.2 Hard Red Winter, f.o.b. was as much as 80 percent higher than in the corresponding period last year, it averaged USD 302 per tonne in October 2011, up marginally from October 2010. Wheat export prices have fallen sharply from USD 364 per tonne in April and their peak of USD 482 per tonne in March 2008.

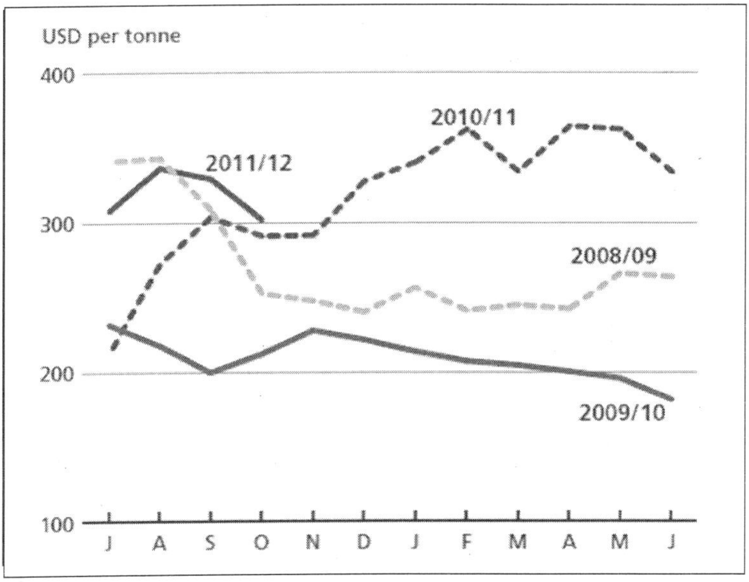

Figure 1. Wheat export price (US no. 2 H.W. Gulf)

However, in spite of a declining price trend, wheat markets remain volatile. International wheat prices continued to move in tandem with swings in maize markets. The much tighter maize balance has been the main driver of price changes in wheat markets since the beginning of the current season, mostly because of the increased use of wheat for animal feed. Furthermore, the fact that maize continues to trade close, if not at a premium, to wheat at the Chicago Board of Trade (CBOT) has become one of the emerging features of the current season.

In October, the CBOT wheat for March 2012 delivery averaged USD 243 per tonne, 9 percent below the corresponding period last year. Larger than expected inventories and generally favourable planting conditions for harvests in 2012 are expected to moderate any upside pressures on wheat prices that might arise. However, amidst the backdrop of much tighter maize markets and macroeconomic uncertainties, wheat prices are likely to remain firm through the remainder of the 2011/12 season.

World Wheat Trade

World wheat trade (exports, including wheat flour in wheat equivalent) in 2011/12 (July/June) is forecast to reach 131 million tonnes, 4.4 percent higher than estimated for 2010/11 and 6 million tonnes more than FAO's first trade forecast for 2011/12, published in the June 2011 *Food Outlook.*

The bulk of the trade expansion in 2011/12 over the previous season is expected to be supported by strong import demand in Europe, primarily in the EU. The EU's total wheat imports are forecast to increase by nearly 3 million tonnes in 2011/12, given this year's small increase in wheat production in the EU but large supplies in the Black Sea region.

In Asia, the biggest wheat importing region, total wheat imports in 2011/12 are forecast to approach 60 million tonnes, up nearly 2 million tonnes from 2010/11. The biggest increases are forecast for Afghanistan, Iraq, Israel, Saudi Arabia, Syrian Arab Republic and Yemen, mostly to compensate for lower domestic outputs, and in China and Indonesia because of a continuing strong growth in domestic demand. However, smaller wheat purchases are forecast for Bangladesh and Turkey, largely because of higher domestic crops, and for Viet Nam, because of relatively large carryover inventories.

Total wheat imports by Africa are forecast to exceed 37 million tonnes, down 500 000 tonnes from 2010/11. The decrease will be mostly due to lower imports by several countries in North Africa, notably Egypt, Morocco and Tunisia, following bigger harvests, more than offsetting larger shipments to several countries in sub-Saharan Africa, particularly Ethiopia, because of lower production and rising domestic prices.

In Latin America and the Caribbean, total wheat imports in 2011/12 are forecast to remain unchanged from the previous season, at around 20.4 million tonnes. Imports by Brazil, the region's largest wheat importer, are forecast to increase slightly, to compensate for a decline in production.

Regarding exports, a sharp recovery in availabilities in the CIS countries is the main feature of the 2011/12 marketing season. In fact, the prospect for exports from the Black Sea region improved steadily as the season progressed and harvests completed, confirming a strong recovery from last year's drought-reduced levels. In the Russian Federation, a strong upturn in production (up 37 percent from last year) and the removal of export restrictions since the start of the season are expected to boost wheat shipments to at least 18.5 million tonnes, close to the record in 2008/09 after the previous season's reduced level. The Government has recently announced that, if total grain exports approach 24 million tonnes, it may impose export taxes. Wheat shipments from Ukraine are forecast to triple, to 9 million tonnes, and a 30 percent growth in exports is expected by Kazakhstan, to 7.2 million tonnes. As a result, the aggregate wheat sales of the three major CIS wheat producers may reach 33 million tonnes, well above the world's largest exporter, the United States. Wheat exports from the United States are forecast to decline this season to 26 million tonnes from 35 million tonnes in 2010/11, due to a cut in domestic production and stronger currency, which reduces its price competitiveness against other export origins. A significant reduction in shipments is also expected in the EU, where a relatively tight domestic supply compared with last year is expected to curb sales by almost one-third to just under 15 million tonnes. However, shipments from the other traditional major exporting countries, namely Argentina, Australia and Canada, are forecast to remain close to the previous season's levels or even rise slightly.

World Wheat Exports

World wheat exports in the 2010/11 marketing year are estimated to be 125,337,000 metric tons. This compares to average World wheat exports of 116,764,000 metric tons since the 1987/88 marketing year, with an average increase of 998,000 metric tons per year over the 24 year period.

The 10 largest World wheat exporting countries or regions shipped an average of 91.6% of World wheat exports over the 1987/88 to 2010/11 period. Over the most recent 3 marketing years, the 5 largest countries in terms of average World wheat exports were the United States (29.00 mmt), the European Union (22.98 mmt), Canada (18.45 mmt), Australia (14.45 mmt) and Russia (14.65 mmt).

The Ukraine, Turkey, Russia and Mexico have the largest historic variability in wheat exports relative to their average export shipments since the 1987/88 marketing year, with cv measures of 0.65, 0.64, 0.62 and 0.50, respectively. Kazakhstan, Argentina, Australia and the European Union, Canada and the United States have lower amounts of relative variability in annual wheat exports. Aggregate World wheat exports have exhibited a relatively low level of variability compared to individual countries since 1987/88. Taken together, these results indicate that annual wheat exports have been more variable in the Ukraine, Turkey, Russia and Mexico than in other major wheat exporting countries and regions. The least amount of annual variability in wheat exports among major wheat exporters is evident in the United States and Canada. Total aggregate World wheat exports have been much less variable than for individual major wheat exporting countries.

World Wheat Imports

World wheat imports in the 2010/11 marketing year are estimated to be 122,779,000 metric tons. This compares to average World wheat imports of 107,238,000 metric tons since the 1987/88 marketing year, with average increases of 912,000 metric tons per year over the 24 year period.

The 10 largest World wheat importing regions secured an average of 95.4% of World wheat imports over the 1987/88 to 2010/11 marketing year period. Over the 3 most recent marketing years, the 6 largest regions in terms of average World wheat imports were North Africa (22.4 mmt), the Middle East (22.2 mmt), Sub Saharan Africa (14.6 mmt), Southeast Asia (13.0 mmt), South America (13.0 mmt), and East Asia (12.5 mmt). East Asia includes China, while Southeast Asia includes India. Turkey is included in the Middle East while Egypt is included in North Africa.

The top 12 wheat importing countries during the 3 year MY 2008/09 through MY 2010/11 period are shown in Chart 14. Over the last three completed marketing years, the #1 wheat importing country was Egypt (averaging 10.0 mmt per marketing year), followed by 2) Brazil (6.5 mmt per MY), 3) the European Union (5.9 mmt), 4) Algeria (5.6 mmt), 5) Indonesia (5.4 mmt), 6) Japan (5.3 mmt), 7) South Korea (3.9 mmt), 8) Nigeria (3.9 mmt), 9) Iraq (3.8 mmt), 10) Mexico (3.3 mmt), 11) Morocco (3.2 mmt) and Bangladesh (3.1 mmt).

The Former Soviet Union region has the largest historic variability in wheat imports relative to their average import shipments since 1987/88, with a cv measure of 0.72. North America and SubSaharan Africa and South Asia, have the next highest levels of relative variation in wheat imports over time, followed by the European Union and the Middle East, East Asia, and Southeast Asia. South America and North Africa have the lowest levels of relative annual variability in wheat imports among major regions of the World. Aggregate World wheat imports have exhibited a relatively low level of variability compared to individual regions since 1987/88, displaying a tendency for offsetting changes to occur across major wheat importing regions of the world.

Among the top 12 wheat importing countries, Nigeria has the largest historic variability in wheat imports relative to their average import shipments since 1987/88, with a cv measure of 0.73, followed by Iraq and Mexico. These major wheat importing countries were followed in relative annual wheat import

variability by Bangladesh, Morocco and Brazil, Indonesia and the European Union (0.32). Compared to the major wheat importing countries, the relative variation in wheat imports by Egypt and Algeria and Japan is markedly lower. Aggregate World wheat imports have exhibited a relatively low level of variability compared to individual regions since 1987/88, displaying a tendency for offsetting changes to occur across major wheat importing countries of the world.

World Wheat Total Use

World wheat total use in the 2010/11 marketing year is estimated to be 788,011,000 metric tons. This compares to average World wheat use of 735,475,000 metric tons since the 1987/88 marketing year, with average increases of 6,397,000 metric tons per year over the 24 year period.

The 20 largest World wheat usage countries used an average of 85.8% of World wheat over the 1987/88 to 2010/11 marketing year period. Over the 3 most recent marketing years, the 6 largest regions in terms of average World wheat usage were the European Union (147.5 mmt & 19% of total usage), China (108.0 mmt & 14% of total use), India (77.3 mmt & 10%), the United States (61.4 mmt & 8%), Russian (56.5 mmt & 7%), and Canada (26.2 mmt & 3%).

Among the top 20 wheat usage countries, the Ukraine has the largest historic variability in total wheat use relative to their average usage since 1987/88, with a cv measure of 0.29, followed by Uzbekistan, Kazakhstan, Australia and Argentina. These major wheat using countries were followed in relative annual wheat import variability by Egypt, Morocco and Russia, Canada, Brazil and Mexico, Pakistan, India, Iran, Turkey, the United States the European Union, Japan and China.

There is a marked distinction between the relative variability of wheat usage in more variable countries such as the Ukraine, Kazakstan, Australia and Argentina, and less variable nations such as China, Japan, the European Union and the United states. This variability among countries depend on wheat production and export prospects for these countries on the one hand and the inelasticity of wheat demand for food and other uses by each country on the other. Aggregate World wheat total use exhibited a relatively low level of variability compared to individual countries since 1987/88, displaying a tendency for offsetting changes in usage to occur across major wheat using countries.

Higher Feed Use Boosts World Wheat Utilization in 2011/12

Contrary to the forecast at the start of the season, for a slight expansion in total wheat utilization in 2011/12, the latest indications point to a significant increase, mostly in response to larger availabilities than anticipated earlier and competitive feed wheat relative to maize. At the current forecast level of 682 million tonnes, total wheat utilization would be 2.2 percent higher than in 2010/11. The 2.2 percent expansion exceeds the ten-year trend as well as the 1.5 percent and 1.8 percent increases in 2010/11 and 2009/10 respectively but lower than the 2.9 percent expansion in 2008/09.

World utilization of wheat for direct human consumption is forecast at 474 million tonnes, up 1 percent from 2010/11. This implies a 68 kg annual consumption per person, which is similar to the levels of the past two seasons, thus indicating that the global food use of wheat is continuing to keep up with average world population growth. Among the most populous countries, per capita wheat consumption in China is likely to continue a slow decline, at just under 64 kg, compared with 73 kg at the start of the millennium. The gradual fall in per capita wheat consumption in China is mainly due to larger intake of more value-added food products. However, in India, per capita

consumption of wheat has been rising over the past decade and is now over 61 kg. While unexpected drops in year-on-year consumption may be rare, they can happen as a result of conflicts, unexpected production and/or economic problems. In 2011/12, the largest consumption declines among major wheat consuming countries are anticipated in Libya with a fall of 4 kg, and in Bahrain and Oman, which will have declines of 2–3 kg.

The latest forecast for total feed utilization of wheat stands at 131 million tonnes, 5.6 percent higher than the revised 2010/11 estimate and up 4 million tonnes from FAO's first forecast in June. As a share of total cereal utilization, feed usage of wheat remains relatively limited and concentrated in few countries or regions with the EU being the largest market for feed wheat. However, in recent years, the usage of wheat for animal feed has been increasing due to its price competiveness compared with coarse grains. In 2011/12, the strong recovery in wheat production in the CIS and larger global supplies of feed wheat, in the face of very tight maize availability, are among the main factors for the surge in wheat usage for animal feed. The fastest expansions are expected in China, the EU and the United States.

The other uses of wheat include industrial use and seeds. The combination of other uses and post harvest losses account for 11 percent of world wheat production in 2011/12. The industrial use of wheat has expanded over the past decade, mostly driven by larger utilization of wheat as feedstock for biofuels (ethanol). According to the International Grains Council (IGC), total wheat used for industrial production in 2011/12 could reach 21 million tonnes, 1 million tonnes higher than in 2010/11. Starch manufacturing constitutes the primary industrial use of wheat followed by biofuels. The IGC forecasts that wheat used for production of biofuels (excluding non-fuel uses) will reach 7.3 million tonnes in 2011/12, 22 percent higher than in 2010/11 mostly is concentrated in the EU (5.3 million tonnes), followed by Canada (2.1 million tonnes) and China (1 million tonnes).

World Wheat Ending Stocks

Total World wheat ending stocks in the 2010/11 marketing year are estimated to be 177,774,000 metric tons. This compares to average World wheat ending stocks of 168,482,000 metric tons since the 1987/88 marketing year, with average increases of 174,000 metric tons per year over the 24 year period.

The 10 largest World wheat countries or regions in terms of ending stocks had an average of 80.7% of total World wheat stocks over the 1987/88 to 2010/11 period of marketing years.

Over the most recent 3 years, the 8 largest countries or regions in terms of average World wheat ending stocks were China (53.3 mmt or 29% of total World ending stocks), the United States (22.2 mmt or 12% of total stocks), the European Union (16.1 mmt or 9%), India (14.7 mmt or 8%), Russia (8.8 mmt or 5%), Canada (6.7 mmt or 4%), Egypt (5.4 mmt or 3%), and Australia (5.0 mmt or 3%).

During the 1997/98 through 2000/01 period, China held 47% of the World's ending wheat stocks. This compares to an average of 30% of the World's ending stocks of wheat held by China during the more recent 2007/08 through 2010/11 period. The large holdings of world wheat stocks during the 1997/98 through 2000/01 period indicate that it is not unprecedented for the Chinese to accumulate even larger proportions of the World's wheat ending stocks than they currently have.

Egypt, India, and Russia have the largest historic variability in wheat ending stocks relative to their average stocks since 1987/88, with cv measures of 0.77, 0.72, and 0.67, respectively. Syria and Australia have lower levels of relative variation in wheat ending stocks over time, followed by the United States, China, Canada, the European Union and Iran. Aggregate World wheat ending stocks have exhibited a relatively low level of variability compared to individual countries since the 1987/88 marketing year.

Total World wheat % ending stockstouse in the 2010/11 marketing year is projected to be 22.6 percent. This compares to average World wheat % ending stockstouse average of 24.5% since the 1987/88 marketing year, with average decreases of 0.2% per year over the 24 year period.

FAO's forecasts for ending stocks in 2012 have been revised up steadily since the beginning of the season. The main reason is bigger harvests than earlier anticipated in many countries. The latest forecast for global stocks takes into account recent adjustments to the size of wheat inventories in the

CIS, China, the EU and the United States. Based on the latest production estimates for 2011 and the utilization forecasts for 2011/12, world wheat stocks are likely to approach 190 million tonnes by the close of the crop seasons in 2012, 10 million tonnes, or 4.3 percent, above their opening levels. This forecast is 4 million tonnes higher than was reported in the October issue of FAO's *Crop Prospect and Food Situation.* The latest upward adjustments reflect expectation of larger ending stocks in several countries; namely Australia, Canada, the Russian Federation, and the United States, more than offsetting downward revisions to the previous forecasts for China and the EU.

Based on the current supply and demand estimates, the global stocks-to-use ratio for 2011/12 is expected to reach 28.2 percent, up sharply from 26.7 percent in 2010/11. It would be well above the low of 21.6 percent registered in 2007/08 as well as its five-year average (2004/05–2008/09) of 26.2 percent. Total wheat stocks held by the five traditional exporters (Argentina, Australia, Canada, the EU and the United States) are forecast to increase slightly from their opening levels, to nearly 50 million tonnes. At the current forecast, the ratio of stocks held by the major exporters to their disappearance (i.e. domestic utilization plus exports) is put at 19 percent, up slightly from 17.6 percent in 2010/11 but well above the low of 13.4 percent in 2007/08.

Among the other large stockholders, wheat inventories in China are forecast to remain stable at around 50 million tonnes. Also in India, where this year's wheat production is also at a record, ending stocks are anticipated unchanged, at around 18 million tonnes.

Emerging Trends

Since at least the 1987/88 marketing year global wheat markets have been strongly influenced by a small number of countries that either produce or export large portions of the World's wheat. Global wheat production has been concentrated (84.3%) in the top 10 wheat producing countries and regions of the World since MY 1987/88, including the top 5 of 1) the European Union, 2) China, 3) India, 4) the United States, and 5) Russia. World wheat exports are also strongly influenced by a limited group of countries and regions, with the top 10 accounting for 91.6% of World wheat exports since MY 1987/88. The top 5 World wheat exporters at this time are 1) the United States, 2) the European Union, 3) Canada, 4) Russia, and 5) Australia, followed closely by the Ukraine, Kazakhstan, and Argentina.

World wheat import markets are less concentrated than World wheat production and export markets in terms of predominant countries or regions. The top 12 world wheat importing countries over the last 10 years have accounted for approximately half of World wheat imports. These supplydemand statistics indicate that fairly concentrated production and export sectors exist in the World wheat market. They also indicate that World wheat import buyers are less concentrated as a group than are World wheat exporters. A relatively small number of wheat export sellers interact with a somewhat larger group of World wheat import buyers in today's wheat market. Two major wheat producing and usage countries (i.e., China and India) focus on domestic usage of wheat, and at this time are unlikely to sell their domestic wheat ending stocks into World wheat export markets. From the broader perspective of economic theory regarding competitive agricultural markets, a relatively large number of countries are competitively involved in buying wheat for imports to help meet the dietary needs of their respective populations, whereas a relatively small number of countries are capable of being major wheat exporters, and are competitively involved as sellers in World wheat export markets.

Ending stocks of wheat in World markets tend to be concentrated in the hands of a limited number of countries or regions. Over the last 10 years, the top 10 wheat stocksowning countries or regions have had ownership of 77.1% of total World wheat ending stocks. These countries include China, the United States, India, the European Union, Canada, Australia and Russia. Eqypt, Iran and Syria have also consistently been among the top 10 holders of World Wheat ending stocks over the last decade.

Because of the availability of natural agricultural farmland resources and the likelihood of continued population growth in the future, the roles of various countries or regions in determination of World wheat market supply demand balances is unlikely to change appreciably in the foreseeable future. However, wheat acreage in the United States has been declining since 1981 due to the profitability of other competitive crops and other factors.

Wheat yield trends in the United States have been relatively flat or only rising slowly over this same time period. The possibility of future declines in U.S. wheat acreage and production, and of declining U.S. wheat exports as well, are among the factors likely to impact how competitive a role the United States will play in World wheat export markets in the future.

References

Briggle, L.W. & Curtis, B.C. (1987). Wheat worlwide. *In* E.G. Heyne, ed. *Wheat and wheat improvement*, 2nd ed. Agronomy 13: 1-32.

Hanson, H., Borlaug, N.E. & Anderson, R.G. (1982). *Wheat in the third world.* Boulder, CO, USA, Westview Press.

Kohli, M.M. & McMahon, M.A. (1988). A perspective of research needs for non-irrigated tropical countries. *In* A.R. Klatt, ed. *Wheat production constraints in tropical environments*. Mexico, DF, CIMMYT.

Leonard, W.H. & Martin, J.H. (1963). *Cereal crops*. New York, NY, USA, MacMillan Publishing.

Peterson, R.F. (1965). *Wheat botany, cultivation and utilization.* New York, NY, USA, Interscience Publishers.

Orth, R.A. & Shellenberger, J.A. (1988). Origin, production, and utilisation of wheat. *In* Y. Pomeranz, ed. *Wheat chemistry and technology*, vol. 3. St Paul, MN, USA, American Association of Cereal Chemists.

9

Foodgrain Market and Global Food Security

By 2050 it is estimated that the world will need to increase food production by 70 percent to feed a larger, more urban, and, it is hoped, wealthier population. It will have to accomplish this in the context of shrinking availability of arable land and water and other environmental constraints. Increased investment in research and development and improvements to production capacity are absolutely vital to increase the *availability* of food. Usually less emphasized but equally crucial for boosting productivity and availability of food is the existence of markets. Markets and trade also contribute to achieving global food security by increasing *access* to food.

This report briefly outlines the food security challenge facing the world and then expands on the often-overlooked role that markets and trade play in achieving food security. Recognizing the important U.S. initiatives under way, the report offers a set of recommendations on how the United States can concretely work to strengthen markets and trade.

The Challenges Ahead

The United Nations Population Division projects the world population to increase from approximately 6.8 billion in 2009 to more than 9.1 billion in 2050. The majority of the growth will take place in developing countries, where population is projected to grow from 5.6 billion in 2009 to 7.9 billion in 2050. Africa's population is expected to double in size between 2009 and 2050.

The world will also undergo rapid urbanization. Between 2007 and 2050, the world urban population is anticipated to increase from 3.3 billion to 6.4 billion. By 2050, it is likely that some 70 percent of the world's population will live in urban areas.

A growing and more urban population will be accompanied by increases in income levels and changing food preferences. As incomes rise from about $2 per day to $10 per day, people prefer more dairy, meat, fruits, and vegetables, requiring an increased production of agricultural commodities, whereas people earning more than $10 per day consume more processed foods, services, packaging, and variety and luxury forms of food, but do not require more agricultural commodities. The Food and Agriculture Organization (FAO) of the United Nations projects that annual cereal production will need to rise from 2.1 billion tons today to about 3 billion tons in 2050 and that annual meat production will need to rise by more than 200 million tons to reach 470 million tons.

Arable land and water constraints, combined with the likely impacts of climate change—more severe weather events, an increase in pests, and longer-term changes to agricultural production due to increasing temperatures—pose additional challenges to feeding a larger population. When a potential contribution of CO_2 fertilization is discounted, most crop yields are projected to decline significantly by 2050 as a result of climate change. Moreover, the agricultural sector contributes some 14 percent of total greenhouse gas emissions (which is comparable to emissions from the transport sector) and may be called upon to increase productivity without increasing, or even reducing, greenhouse gas emissions.

Self-sufficiency in food production is not a viable solution for most countries, as it is neither economically nor environmentally sustainable. With daunting challenges on the horizon, countries will need to rely more heavily on trade to be food secure. However, there are persistent trade and market access obstacles—in both high-income and developing countries—that impede efficient distribution of, and access to, food.

Role of Markets and Trade

In this section, which speaks to the importance of markets and trade in achieving food security, we emphasize the importance of both "hard" and "soft" infrastructure in creating sound markets.

Markets are necessary to boost productivity and availability. Improved access to agricultural input markets—such as seed and fertilizer—is crucial for productivity growth. Moreover, farmers will only increase production if they have access to viable markets for their agricultural outputs.

Markets and trade also play a crucial role in achieving global food security by increasing access to food. At the simplest level, trade allows food to flow from areas of surplus to areas of deficit—in local, regional, and global markets. Well-functioning markets transmit price signals, which allow changes in demand to be met by supply. When demand is greater than supply, producers increase production in response to price signals, and this increased production, in turn, helps to stabilize prices. By transmitting information in this way, markets help to reduce price volatility. As witnessed during the 2007–2008 spike in food prices, it is the poor who suffer the most from volatility, as they spend proportionately a much larger share of their income on food. For example, in Bangladesh, expenditures for rice alone account for 48 percent of total food expenditures and 30 percent of total household expenditures.

In addition to effectively allocating and distributing the world's food supply, improved access to agricultural markets also serves as a powerful tool for economic growth in less-developed countries. In regions like sub-Saharan Africa, where 70 percent of the population relies on agriculture for their livelihood, and 80 percent of all the farms are less than 2 hectares in size, poor smallholder farmers can turn their surpluses into income only if they have the ability to access markets. Increased incomes, in turn, increase food security and help to alleviate poverty.

To enhance existing markets and create new ones requires addressing the serious "supply-side constraints" that face many developing countries. These constraints include missing or inadequate physical infrastructure, such as roads and ports and storage and processing facilities. The lack of rural and regional infrastructure to connect neighbouring countries inhibits the ability of rural farmers to receive necessary inputs such as seeds, fertilizer, pesticides, and agricultural equipment and their ability to move their products on to local and regional markets. The increased transaction costs do not allow producers to benefit from economies of scale. Physical infrastructure aside, a weak legal and policy environment can also constrain private sector activity. Governments should seek to provide a sound enabling environment for the private sector that gives utmost importance to establishing the rule of law, including contract enforcement procedures, a sound financial system, and clearly defined property rights.

The initiative for tackling supply-side constraints in developing countries needs to be taken by the governments of those countries, with ongoing input from the private sector. Cross border constraints need to be tackled on a regional level. Developed countries, however, can also play an important role. Not only can they help address supply-side constraints through financial assistance, but they can also help create positive production incentives by reforming the international agricultural trade system and their own domestic policies.

Global Grain Contract: A New Food Security Instrument

When futures markets sprung to life in Chicago about 150 years ago, no one could have envisaged how colossal they would become in the twenty-first century. Established as a clubhouse and insulated from public view, futures markets today have achieved celebrity status. Growing at remarkable rates for the past several years, these markets now attract interest from governments, the media, the financial world and the general public. However, outside the circle of professional users, futures markets in general are poorly understood and most recently have been lumped together with other sorts of "investments." This misunderstanding may be masking a potential beneficial role for futures in global price discovery.

Futures contracts are unique instruments. Although various writers have recently classified futures as a type of security, from both a legal and operational viewpoint, they do not meet the securities designation. While securities - such as equity shares or bonds - are issued under strict legal standards by corporate or governmental entities, futures are purchase and sales agreements created by an exchange. Also, all securities offerings require United States Securities and Exchange Commission (SEC) registration (or other national supervisory body) and strict disclosure documents describing the issuing entity - *viz.*- business model, operations, financial results, management structure, etc. whereas futures contracts are filed with the United States Commodity Futures Trading Commission (CFTC) determines as a "Self Certification Submission." If the CFTC that the contract complies with the Commodity Exchange Act, particularly with regard to anti-manipulation provisions, it will allow the petitioning exchange to list the contract for trading.

Additionally, different regulations govern futures and securities. Most notably, rules prohibiting insider trading in securities do not apply to futures trading. As those inside a company play a fiduciary role to shareholders, they are prohibited from disclosing or trading on material non-public information regarding such matters as a takeovers or trade secrets. The concept of fiduciary does not pertain to commodity futures transactions - commodity traders possess varying amounts of information about crops or markets such as weather events, export sales, or government interventions, but there are no rules regulating the flow of such information. An exporter can disclose knowledge about potential export business or recent trading activity to anyone. Also many commodity related services publish daily reports on cash transactions and transportation rates. It is, however, illegal for a government official to leak knowledge of an official United States Department of Agriculture (USDA) crop report prior to its scheduled announcement.

The regulatory frameworks of securities and futures have many similarities, such as prohibitions against fraud, or front running, but otherwise have different objectives. Investor protection is the central aim of securities regulation, while preventing market manipulation is the intent of futures regulation. Both the exchange and the CFTC monitor the cash market and actively dissuade traders ("jawboning") from such practices as physical hoarding or non-economic movement of the commodity (distortion of trade). They also oversee the futures trading to guard against corners, squeezes or other price distorting activities.

Operationally, futures and securities differ sharply; futures contracts are hedging vehicles for the purpose of transferring risk - listed in serial months to suit hedging needs. Securities are investment instruments; equity securities represent a share of ownership in a corporation and bonds represent a creditor relationship between bondholder and issuer. In futures, speculators - i.e. traders that have no commercial interest in the underlying asset of the futures contract - hope to profit from taking the opposite side of hedgers' buy and sell orders. In securities markets, the securities buyer hopes to profit from share price increases, dividends or interest payments. In both markets, the speculator and the investor face risk of monetary loss, but from different circumstances. The speculator's loss arises solely from price risk exposure - which may be substantial. The investor's loss, which can stem from declining securities' prices, may arise also from debt default or bankruptcy of the issuing entity - the latter causing forfeiture of the entire investment.

Finally, while securities markets rely on a depository system to settle changes in ownership, futures markets interpose a central counterparty called the clearinghouse between every transaction. The clearinghouse - as the buyer to every seller and the seller to every buyer, relies on a margining system to eliminate the risk of default among market participants. To initiate trading, every clearing member must deposit with the clearinghouse an initial margin, which acts as a performance bond - usually equal to 5-10 percent of contract value. Because the clearinghouse "marks-to-market" members' trading positions, it collects and remits margin monies on a daily basis in accordance with the profit or loss on positions held or closed out. Significantly, gains and losses always offset each other and the number of buy and sell contracts always match. Hence, futures trading is called a zero sum game.

Futures contracts are standardized. For commodity contracts, the terms - usually called the specifications - will include underlying asset, contract size (by weight or volume), quality, currency denomination, minimum price fluctuation, delivery location or pricing basis and method of delivery or settlement. Other terms may include differentials for quality variations or for alternate delivery locations. In some contracts the final expiration price is cash settled, meaning that the price is derived from a formula of cash prices usually reported by reputable cash dealers. Maximum daily price limits and maximum position limits are also usually included in contract terms. Exchanges commonly create rules for dealing with potential price congestion or manipulation, default, and *force majeure*.

In the United States of America, where commodity futures contracts have existed the longest, futures contracts originally were constructed by the major commercial players. Since the United States of America became a major exporter of wheat starting in the middle of the 19th century with Chicago as the primary storage and trans-shipment hub, the contracts were meant to be most useful to warehouses which bought cheap cash wheat from the farmer at harvest and simultaneously sold a higher priced deferred futures month when shortages would likely arise. By doing so, the warehouse would "lock in" a profit that would more than offset its cost of storing and handling the grain. Because the warehouse controlled the stocks and the issuance of warehouse receipts - which became legal instruments in the state of Illinois in 1871 - it had considerable pricing power over the market. For example - only the warehouse with graded stocks (and therefore warehouse receipts) in its silos could establish a short futures position and then make delivery against the contract. Other short sellers would have to buy back their short positions, even if the futures price were a substantial premium to the physical commodity that could be brought alongside the elevator. In other words, farmers or grain operators would always take a discount to the futures price from the Chicago warehouse, as they had no way of turning their goods into registered warehouse receipts.

Table 1: Securities versus futures

Securities	*Commodity futures*
Corporation/government issued	Exchange created
SEC registration	CFTC submission
Disclosure requirements	CEA compliance
Insider trading prohibition	Free information fl ow
Investor protections	Anti-manipulation provisions
Investment instruments	Risk transfer vehicles
Mostly collateralized	Highly leveraged
Intangible property	Commodity delivery obligations
Depository settlement system	Central clearinghouse
Capital formation	Zero sum game

Nonetheless, despite the power of the warehouse to control much of the delivery situation on the short side, long speculators could and often did acquire long positions exceeding the warehouse's abilities to accumulate grain and make delivery, especially in times of shortages. Corners and squeezes were a frequent event at the Chicago Board of Trade from its inception, sometimes involving the shipment of foreign wheat into the terminal area to break the corner. Finally, after WWI the Government of the United States of America put various measures in place to regulate and monitor the trading of futures contracts in agricultural commodities.

Mechanics of Futures and Hedging

As futures contracts approach expiration their prices begin to converge with underlying cash values, specified as the pricing basis of the contract. For example, the CBOT soybean contract specifies delivery of soybeans to be loaded into barges along the Illinois River. So, theoretically, the futures price will rise or decline to approximate the fob barge price during the delivery period. If futures prices are too high relative to cash, barge loading stations will tend to deliver against short futures sales, as the high futures price will represent a better price for their soybeans than the cash market. Conversely, if futures prices are too low relative to cash, long holders will tend to maintain their long positions to obtain the delivered commodity or until the price rises sufficiently to approximate cash values.

In the Chicago futures system, deliveries are tendered by the short to the clearinghouse which then assigns them to the long with the "oldest" purchase date. Although some futures contracts (such as the Zhengzhou cotton contract) permit the long to state some preference on quality or delivery locations with the clearinghouse, most futures contracts leave all options to the short. Hence, the least valued quality at the least desirable location will tend to be delivered first. Deliveries are tendered and stopped throughout the delivery month meaning that the same delivery instruments (in this case shipping certificates as opposed to warehouse receipts) can be "issued" and "stopped" multiple times by multiple players as futures trading continues from the first day of the delivery month until mid-month. Other countries, such as India, do not allow retendering. There, after contract expiration, the clearinghouse compels the outstanding shorts to tender their deliveries which it then assigns to the outstanding longs. The delivery process is the key in bringing proper convergence between futures and cash - so that the basis level around the delivery market should be close to zero at contract expiration.

Hedging in commodity futures markets has been recognized as proper business activity from the market's inception. Well functioning future markets with ample long and short hedging orders tend to lessen volatility and reduce the trough to peak pricing inherent in the crop cycle. In markets without futures pricing, distressed harvest selling by farmers and end-of-year price spikes by end-users usually characterize price behaviour. Because commodities are volatile, the purpose of hedging is to diminish the price risk of a forward sale or purchase in the physical market. Futures can be thought of as proxy instruments to be held in place until the real transaction can occur: at this point, the hedger will offset the hedge by buying back its futures short or selling out its futures long.

Standard short hedge

Producers are standard short hedgers. After springtime crop sowing, the producer can sell a quantity of futures "short" against anticipated crop production. The sale will protect the producer against falling prices after the crop is harvested during fall. At harvest time, while making a cash market sale, the producer will buy back the hedge (same quantity/same contract month), thereby offsetting the trade. In the simplified example below ("zero" basis assumption), the producer anticipates a production of 50 000 bushels of corn - which is the equivalent of 10 futures contracts - 5 000 bushels each. In June, the producer decides to sell 10 December futures contracts at USD 5.00 and later, during November, buys back the futures at USD 4.00 while selling the harvested production at USD 4.00. The producer realizes USD 5.00 per bushel for the sale. Whether the price increases or decreases, the producer would realize the same USD 5.00 for the grain by executing the USD 5.00 hedge.

Standard long hedger

Wheat millers are standard long hedgers. A wheat miller can protect against rising prices of wheat by buying futures contracts equal to its milling needs for a particular time period. In this example, the miller determines in September that it can profitably mill wheat into flour at the prevailing USD 6.00 per bushel January cash price and buys 100 contracts of wheat at USD 6.75 (minus USD 0.75 basis assumption). By January, the wheat price has dropped unexpectedly allowing the miller to buy cash wheat USD 1.00 lower, at USD 5.00. The basis level, however, has appreciated because farmers are reluctant to sell after the price drop and is now minus USD 0.25. After buying the USD 5.00 cash wheat, the miller sells out futures long hedge at USD 5.25, incurring a futures loss of USD 1.50. The miller's cost of wheat is therefore USD 0.50 higher than the "locked in" price of USD 6.00 making the purchase price USD 6.50.

Pitfalls of hedging

The standard hedges in the previous paragraphs in actual practice may turn out very differently depending upon developments in the crop year. The producer may only harvest half a crop for example, meaning that the loss incurred by buying back the hedge could dramatically diminish the proceeds from the cash sale. The following example illustrates the outcome of a producer using futures to hedge 100 000 bushels of soybean production at USD 10.00 - anticipating "locked in" revenue of USD 1 000 000 with devastating consequences when drought cuts harvest to 50 000 bushels. In addition to quantity mismatches, other factors can undermine hedging strategies and execution.

Quality mismatch

The particular attributes of a commodity can make a significant difference in its value. For example, in 2008, wheat millers which hedged their needs for dark northern spring wheat in the CBOT soft red winter contract, would have lost over USD 10 per bushel as DNS wheat traded as high as USD 25 while the Soft Red Winter (SRW) wheat contract reached USD 12.

Timing Mismatch Combined With Extreme Backwardation or Contango

In 1996, grain companies in the United States of America promoted a special type of hedging contract to producers called "hedge-to-arrive." Producers were urged to hedge expected harvest production in the July contract rather than the December because of the approximate USD 0.70 per bushel premium of the July over the December contract, a structure called "backwardation." For example, the July and December contract were trading around USD 3.70 and USD 3.00 respectively. The July hedge would eventually have to be "rolled" into the December to actually fix the cash price. As the crop year proceeded the July/December differential (called a "spread") widened substantially to about USD 1.60 - premium July. Producers that had to roll the July into the December (by buying back the July short and selling the December) incurred up to USD 1.60 per bushel loss in executing this strategy, cutting their cash price in half. A similar loss can occur when long hedgers need to roll hedges forward (selling spot and buying deferred contracts) and the market is configured in a steep contango - i.e. spot month is heavily discounted to deferred.

Basis Trading and Basis Risk

The 'basis', or the differential between cash and futures prices, is a common trading vehicle among commercial traders. The phenomenon of basis trading was first documented in the 1950s by Holbrooke Working. Taking issue with Keynes concept of natural backwardation, Working proposed that hedgers used futures not as a risk aversion strategy but as a means of maximizing profits by trading the basis. It is usually more predictable, especially for exporters with extensive logistical capacity and knowledge of global and regional supply and demand fundamentals. Theoretically, the basis cannot trade higher than the futures price plus the full cost of shipping the commodity to another site. If the basis approaches this value, then traders will simply buy the futures and use the deliveries as a source of cash to satisfy a short position elsewhere. For example, if the corn cash basis is trading at a USD 1.00 premium to the March futures for Gulf fob cargos for April shipment and the costs of barge shipment and fobbing the corn into a vessel is USD 0.95, then traders will sell the gulf basis and maintain a long March futures position until they receive delivery. This strategy can create considerable profit for a firm possessing shipping and fobbing capacity: as corn is drawn out of the delivery market, typically the futures price will rise; and, when the barge-loads of corn reach the gulf export market, the exporter can profit from both sides of the trade.

The basis, however, can present tremendous risk - particularly to a hedger, often called an "out of position" hedger. An out of position hedger is a trader long or short a commodity in a region that is distant from the price basis of the futures contract. For example, in 2007, if owners of wheat in the Black Sea region or South Asian wheat growing regions sold CBOT futures as a hedge against that ownership, they would have incurred significant losses. As export taxes (or outright export bans) kept wheat prices in the region artificially low, the price of CBOT wheat tripled - from around USD 4.00 to USD 12.00. The CBOT price spike was a response to a global supply shock and an accompanying rise in demand for United States of America origin wheat. Because of the asymmetrical pricing between the two regions, the futures loss could not have been offset by gains in cash wheat ownership.

Another instance of problems associated with "out of position" hedging occurred in July 2010 on the Euronext-Liffe cocoa contract when a single hedge fund purportedly took delivery of virtually all of the cocoa tonnage in the delivery markets and sent the cocoa price to a record level. Because the delivery markets are in Northern European ports, such as Amsterdam, Antwerp and Hamburg, the growers of cocoa in the African countries of Cote d'lvoire and Ghana could not sell futures against production and then make delivery to profit from the high price. Indeed, any short in the July contract without registered delivery warehouse capacity was forced to buy back its short at a considerable loss. Unlike the CBOT wheat case, the price spike could be mostly attributed to the activities of a single player -after the July futures expired, the September futures dropped by almost 30 percent.

Domestic initiatives have been extremely successful in helping the producer side of the market. Many recent studies have demonstrated how futures have aided producer income realization, by reducing distressed selling, facilitating credit, offering marketing choices, and rewarding quality production. However, government policy constrains most domestic markets. The Government of India for example halted trading in several basic food commodities after perceiving inflation in wheat, tur, dal, potatoes and sugar. The Government of China has frequently intervened in futures markets. Consequently, futures in emerging markets have limited suitability for international hedging and in some cases (e.g. India) foreign direct investment is prohibited. On the opposite side of the supply chain, import dependent countries have almost no instruments to manage commodity price risk outside the international markets such as the CBOT and more recently the Euronext Liffe, where the milling wheat, corn and rapeseed contracts have been gaining benchmark status rapidly. Perhaps now is the time to explore new instruments for addressing the consumption side of the market.

Global Contracts

Because of their unique attributes, futures contracts are open to a wide range of design options. Past futures contract creation favoured warehouses and commercial exporters. With few exceptions, these contracts were based on a single country of origin. The large international exchanges could, however, construct global contracts for cereal and oilseed markets that would complement their current product offerings. Instead of tracking prices that converge with cash values in a single geographic area, global contracts could track "cheapest to deliver" commodities by designating delivery points all over the world. As noted previously, commodities of least value are the ones tendered first by the short. This means, for example, that traders in countries with comparative surpluses and hence low relative prices would deliver on the futures. These deliveries would set the price of the contract.

A precedent for global contracts in the commodity futures market does exist: both the Intercontinental Exchange and Euronext Liffe list a global sugar futures contract. The Euronext Liffe contract - based on white sugar - specifies delivery fob vessel in over three dozen countries. The exchanges designed the contracts as such because of the international structure of sugar production, including its staggered hemispheric growths. The ports able to originate the cheapest sugar (with respect to contract differentials) are the first to deliver against the contract. This unique delivery system is a global signalling system of both price and regional supply availabilities - ready for export, unlike interior based delivery systems that are centred in one geographic location. In addition, such a contract would tend to better absorb events such as export bans or export taxes declared by some countries, as it would spread the price impact of a supply or demand shock across all potential exporting countries.

A global fob contract does have a drawback to potential long takers: the uncertainty of delivery location makes logistical planning - including vessel chartering - complicated and more costly than standard futures. However, this type of drawback may be a perfect antidote to the financialization of futures that has characterized futures markets in the United States of America, evidenced by the build-up of speculative long positions reported in the CFTC Commitment of Traders. Because the delivery-taker would need to charter a vessel and execute a *bona fide* sale to another country, the contract would maintain its integrity to the cash market.

A case for wheat

A global fob wheat contract that would specify fob delivery points in the major producing regions such Australia, Argentina, the Black Sea region, Canada, France and the United States of America would be the most logical initial contract to develop for either cereals or oilseeds. Wheat has multiple origination possibilities and is the most basic food grain shipped in international markets. The contract could also contain contract terms to attract speculators but ensure ultimately that the wheat was channelled into proper commercial channels. In fact, such a contract would attract speculators that arbitrage between two markets in the same commodity (e.g. long French wheat/short global wheat). Arbitragers play an important role in reducing volatility and creating price efficiency. These terms could include:

- Delivery every other month - or alternatively every month to ensure proper convergence with the cash market throughout the year
- Compulsory load-out by the long taker within 60 days of receiving notice of delivery. Compulsory load-out is a feature of Indian futures markets and prevents stocks accumulation by speculators who keep the commodity insulated from commercial channels

- Speculative position limits which are the same for every month, e.g. 1 000 lanuary, 1 000, March, 1 000 May, etc. This would prevent the distorting "roll" before every delivery month
- Contract size denominated in 100 tonnes
- Delivery quantity issued by load-out elevator in multiples of 5 000 tonnes
- Quality to include both hard red wheat and white wheat at differentials

The currency denomination of this contract could involve a hybrid approach of trading in dollars and calculated simultaneously in Special Drawing Rights (SDRs), which are international currency reserve assets issued by International Monetary Fund. SDRs are not a currency *per se* but rather a currency derivative. The SDR value is the reciprocal of the sum of the weighted basket of four currencies (quoted in exchange rate equivalents): the British Pound, the Euro, the Japanese Yen and the United States Dollar. It is recalculated every day. Although intended help countries with reserve issues, SDRs have received endorsement by several economists, notably World Bank President Robert Zoellick, as a possible basis for a new monetary system. Officials in China have also proposed the development of a new monetary system based on SDRs. Finally the IMF in December of 2010 has urged the transition from a single reserve currency (United States Dollar) to SDRs. Although it is doubtful that either SDRs or another currency such as the Euro would start to denominate commodities in the immediate future, a double quote system would offer an alternate pricing view, particularly as many countries that draw on SDRs are commodity importers.

Because SDRs are calculated as a reciprocal to non-United States Dollar currencies - they could provide a price mechanism that smoothes out the volatility of United States Dollar based commodities - which tend to rise when the United States Dollar declines. While the United States Dollar value is always held constant as an exchange rate of 1:1, the other three currency exchange rates are computed against the United States Dollar. If, for example using the chart below, the Euro exchange rate were revalued at 2.00 instead of 1.31 (meaning that the United States Dollar was considerably weaker), then the SDR value would be recomputed (rounded to two digits) from 0.65 at 0.55. If United States Dollar based commodities rose or fell based on weakness or strength, the SDR value would move in opposite direction of commodity price. For example, a wheat price of USD

300 per tonne would be calculated at around 196 SDRs with the SDR rate at 0.65. If wheat rose to USD 360 per tonne and the SDR rate fell to 0.55 due to a weaker United States Dollar, then wheat would be only slightly higher in SDRs - or 198 SDRs.

In sum, the world needs greater understanding of the characteristics, role and possibilities of futures markets in today's globalized environment. Although futures markets have experienced phenomenal growth worldwide over the past ten years, the current system appears insufficient to serve "out of position hedgers" (long cash/short futures) and commodity importers (short cash/long futures). A global contract with multiple delivery ports containing safeguards against excessive speculation and assurances of commercial viability could help remedy the current market shortcomings. A hybrid quote system of dollars and SDRs could prove to be an interesting test case for commodity pricing.

Strengthening Global Food Market Monitoring

Sharp price rises of basic food commodities on world markets in recent years have been the subject of intense interest and concern on the part of consumers and importers everywhere. Doubts have been voiced about whether international markets can be relied upon to meet importing countries' needs, and, for this reason, there is concern over food security implications. Over the long-run, periods of high world prices have been less frequent than declines - roughly one year in four in the case of cereals, and something similar in the case of oilseeds, oils and meals. This helps to explain why the bulk of both international and national efforts have been directed at addressing problems of low, rather than high, prices. Thus, over the years, enormous sums have been spent on buttressing farm incomes largely in richer countries, while far smaller sums on keeping consumer prices at accessible levels mainly in poorer countries. Reflecting this imbalance, most of the trade liberalization negotiations have f ocussed on reducing and constraining support to producers. Liberalization was widely expected to lower price variability even though, at the same time, a reduction in carryovers was also expected, which on its own would act to increase price variability.

Over the past few decades, international efforts to tackle price volatility focused on two main approaches:

- the attempt to negotiate international commodity agreements to stabilize prices, which saw the greatest thrust in the 1970s, after which they have been rather low-key; and
- the development of various compensatory financing arrangements, that is the International Monetary Fund's Compensatory Financing Facility (CFF) - which have provided some help to eligible countries, but nothing in any way comparable to producer support in many Organisation for Economic Co-operation and Development (OECD) countries.

Currently, there is growing interest in the methods that might be used to limit the impact of non-commercial financial sector investment in futures contracts for food commodities. Whatever the merits of current proposals, it is clear that this type of approach could affect price volatility and perhaps serve to reduce extreme fluctuations. In this sense, moves to regulate futures markets are akin to the earlier approaches to international commodity agreements that operated on the physical volumes traded so as to help stabilize prices in the market.

The basic causes of periods of sharply rising market prices have varied somewhat over the years.

In the mid-1970s, during the world food crisis, grain stocks had been very low, prices of petroleum and inputs had been rising beforehand, when a sharp rise in import demand by the former USSR sent grains prices soaring. In the case of rice stocks, which were also low, a very thin world market was not able to handle the sudden surge in import demand that followed a poor monsoon that afflicted the rice-growing belt of South Asia. Some exporters introduced export taxes, and there was a brief export ban on oilseeds by the world's largest exporter, the United States of America.

In the mid-1990s, while stocks had been run down as part of the policy to curb public support of agriculture during the liberalization process, the deterioration of the global supply/demand situation caused world prices to soar.

In the 2008-2010 double-spike of cereals and oilseeds prices, which also affected livestock and dairy products, the main causes were the low level of exportable supplies in a period of rising utilization (in part owing to sharply increased use of grains and oil crops as biofuel, as well as a welcome rise in consumption in some rapidly expanding developing economies, particularly in Asia). For some cereals and oilseeds, the spikes were aggravated by substantial purchases of futures contracts by non-commercial agents and by national trade policies, including temporary export bans that limited the market response of a number of countries.

In 2010 there was an unexpected weather-driven drop in grain production in some key exporting countries together with restrictions on exports by some.

Growing Demand for more Accurate and Timely Market Information

In September 2010, the one-day Extraordinary Intersessional Meeting of the FAO's Intergovernmental Group (IGG) on Grains and Rice noted that among the root causes of recent price volatility was the lack of reliable and up-to-date information on crop supply and demand and export availability. Therefore, the IGG recommended that the FAO intensify its information gathering and dissemination at all levels (FAO, 2010). While this recommendation was addressed to the cereals sector, there is little doubt that the same can be said of the other major foodstuffs too. The problem is widespread. Despite the increase in the volume of raw data and the greater speed of transmitting information over recent years, the capacity to analyse the mass of often conflicting and variable quality data and to disseminate the resulting analyses has not kept pace particularly in the public, free-access sector.

At the national level, the capacity of many countries to collect and process basic agricultural data has often deteriorated, and public statistical services have difficulties undertaking such forward-looking exercises as crop forecasts, let alone comprehensive supply/demand analysis and trade forecasts. The IGGs recognized this weakness and recommended action to strengthen capacity of all partners "in relation to monitoring planting intensions, crop development and domestic market information". As one of the partners, FAO was also requested to improve its own contribution.

The approach that FAO is pursuing in order to enhance its global monitoring activity is based on the fact that the bulk of world production, consumption, stocks and trade is accounted for by a relatively limited number of countries. A significant improvement in the ability to monitor world food markets will necessarily involve making improvements to this key set of major country/commodity elements. For such important food crops as rice, wheat, coarse grains and soybeans, access to accurate information on production in a few countries can go a long way in helping to understand market trends at the global level. For instance, in the case of wheat and rice, less than ten countries account for over 90 percent of world production. Good information on these countries would alone make for a much-improved picture of the global situation. Of course, it is always desirable to strengthen food monitoring for all countries, but it is felt that the most efficient way to respond to the type of requirements listed by FAO's IGGs is to focus on the main market movers.

Regarding the quality of the short-run supply/demand assessments, information on all of these markets movers has strong points as well as weak ones. Some historical databases, on which forecasts are necessarily founded, are weak; others have highly variable weather patterns, or rely on rain-fed production which makes monitoring particularly difficult; some simply don't publish information on key variables; and others are vast countries with many different crop seasons that make aggregation difficult. In addition, across crops, planting and harvesting periods are often very different in most countries. Given a limited potential for expanding total agricultural land in the short-run, changes in plantings of one crop can influence the size of land dedicated to other crops, which is another important factor that will require closer monitoring.

Production Forecasts

It is evident that production forecasts remain at the centre of world food market assessments. Though it has been long perceived to be the main cause of variations in supply and demand balances, rarely has this been so evident as in 2010 when grain production was hit by unexpected weather shocks in several major producing regions almost simultaneously. Moreover, with an increasing proportion of world grain supplies originating from the Black Sea region, an area known for its large variations in yields, unexpected production variations are likely to emerge as a more common feature rather than an exception in the years to come.

In the few months prior to the 2010 price surge, international and national agencies were expecting bumper crops and a generally favourable supply outlook world-wide for the 2010/11 marketing season. The drought-reduced production in the Russian Federation coupled with reduced harvests (also weather related) in other major Commonwealth of Independent States (CIS) producing countries, as well as in Canada and in the European Union, changed the outlook considerably. Events in the Russian Federation, which included repeated downward revisions to production forecasts and the subsequent ban on exports, acted as a leading catalyst for the surge in world price of major grains between late July and mid-August 2010. Given the country's growing importance as a major grain supplier to world markets (the world's fourth largest wheat exporter in 2009/10), a sudden substantial cut in its production or exports was bound to have a major bearing on world markets, as it did in 2010.

Although the first official indication of a major fall in 2010 grain production in the Russian Federation appeared in late July (i.e. few weeks before the harvest), many private agents, both inside and outside the country, were forecasting a fall in output from April. In retrospect it seems that private forecasters were monitoring crop conditions in major growing areas more closely than the public authorities. Because the FAO based its production forecasts for major producing countries on official sources, its earlier forecasts for grain production in the Russian Federation were too high and had to be revised down sharply several times.

A similar situation emerged in the United States of America, this time with regard to maize production. The United States of America is the world's largest producer, user and exporter of maize. For this reason, the maize supply and demand balance in the United States of America has a major impact on world maize markets. The early expectation for the 2010 maize crop in the United States of America pointed to an increase in output (from 2009) to a near record level. Instead, as the season progressed, unfavourable weather conditions (too much rain) hampered yields, reduced production prospects and eventually resulted in maize production falling below the 2009 level. In spite of unfavourable weather conditions during the growing season, the official forecast for maize production in 2010 remained high until very near the harvest. Only in early October (i.e. one month before the harvesting period) crop forecasts were revised sharply lower by the United States Department of Agriculture (USDA) in their World Agricultural Supply and Demand Estimates (WASDE) report of October 2010. This late revision contrasted with expectations of private agents, such as traders, investment firms and banks, who were forecasting lower yields (and hence lower production) from August onward. In most cases, private agents utilized the official area estimates published by the USDA but based their production forecasts on their own yield surveys and field observations. For example, a leading trading house reported that its "early and accurate read on the weather, such as a drought in Russia" in the summer 2010 allowed it to anticipate lower crop yields and position its trading strategy accordingly. This resulted in boosting its profit margin and earnings that year.

Improved Production Forecasts

Traditionally, production forecasts are made up of separate forecasts of area and yield. The area planted to an annual crop results from the translation of earlier planting intentions and developments at the outset of the planting period such as delayed rains - a common problem in non-irrigated agriculture and illustrated every year in the time of arrival of the southwest monsoon. Planting intensions are published officially for some crops and by some countries, but by no means for all of them and sometimes the quality of these reports leaves much to be desired.

Private traders have a special interest in production forecasts, and many either circulate their assessments to subscribers to their newsletters or, if contacted, are prepared to share their views with other analysts, often during trade conferences or trade fairs. Subscribing to newsletters, exchanging views and estimates with the private sector and/or attending meetings of traders is a valuable way of supplementing official announcements of planting intensions. Another approach is to arrange for a local consultant to prepare an annual report on the outlook early in the season in each of the key countries. The person should be someone with experience of the sector and with knowledge of the government officers at the Ministry of Agriculture.

For perennial tree crops (palm, olive and coconut), the main factor is not only the area covered, but also the age structure of the trees: the same trees as the year before will have a different potential output because yield varies with the age of the tree. The stock of productive trees can be raised by planting out young trees from nurseries, but this is usually a relatively small factor. Supporting periodic censuses of tree stocks is a great help to making more accurate output forecasts. Indeed, the improvement of baseline production and area data for all crops annuals and perennials is always of use in improving production outlook work. In cases where part of the problem lies with the data for recent years, one of the strategies could well be the mounting of specialized statistical missions to improve local capacity, undertake censuses and generally improve forecasting ability locally.

The area planted lends itself to model building, and huge numbers of agricultural supply studies have been undertaken over the years. Where suitable studies are available, they can be used to make short-run forecasts of area planted. But a word of caution is needed regarding the use of "off-the-peg" models. They have usually been developed with a different purpose from area forecasting - either for long-term projections, policy modelling or for welfare analyses. Rarely will they be easy to "calibrate" to current levels of production and price. It is usually advisable to construct a relatively simple model tailored to answering the immediate needs of short-run forecasting. This type of model needs to include expected prices of the crop concerned, together with prices of the main competing crop(s) and input costs. It should reflect current policy decisions on, for example, government procurement policies, water allocation decisions and input subsidy programmes.

In addition, high and volatile prices of nearly all major food crops, as experienced in recent years, make advance estimation of the eventual size of plantings more difficult. To most farmers, a higher price of one crop is a reason to plant more of that crop. The other reason is the general anticipation that even if prices were to decline, the decrease could be less than those of the other competing crop(s). In other words, farmers normally make their planting decisions based on relative profitability between among crops. However, in countries where several crops with often similar planting periods are grown, farmers may find it less risky to expand production of not one crop but a combination of crops This factor not only complicates the calculation of the extent of plantings but also may lessen the impact of supply response to high prices.

As the growing season progresses, the focus of forecasting switches from area planted to yield expectations. Again, official forecasts are often made by the major producers, but owing to weather variability and the incidence of unexpected pests or disease there is inevitably a degree of error in any forecast. The hope is that many of these factors will, in a large producing country, balance each other, but, because production is often fairly heavily concentrated in just a few most favourable areas, this hope is not always realized. Weather can nowadays be fairly accurately recorded down to small producing areas, but translating current and accumulated rainfall, soil moisture levels, hours of sunshine at critical junctures in the growth of plants, temperatures, snow cover for winter grains, correlation between the weather and pest and disease build up as well as wind and flood damage and human decisions to hoe, to fertilize and to apply pesticides is a very complex matter and, probably, not an efficient way to forecast yields. For tree crops there are additional factors influencing yield; like the on-off yield cycles or prolonged effects of El Nino (and La Nina) weather patterns on palms for 12 months or more. The development of computer modelling will eventually help, but for the present, it is probably better to sample the opinions of farmers as to the state of their crops. This, again, can be done by government agencies, producer associations and private traders, and their opinions can be sought or bought in the same way as forecasts of area planted.

Hiring local experts is a useful approach towards yield monitoring, provided they can frequently visit the main areas at risk of yield variation. Combing local newspapers and monitoring local radio can also yield helpful warnings. In undertaking detailed local monitoring it is essential to have a good baseline of county/region level area, yield and production so that local pieces of information can be translated into effects on eventual production. Baseline data and a survey of local media sources should be undertaken as a one-off piece of research to improve monitoring.

Knowing the Level of Stocks

If production forecasts are the key element in supply/demand forecasts, stocks are the next most sensitive element. Data on end-of-season stocks, the point during the year when they are at their lowest level, are what is required. Figures at any other time represent supply on hand, not carryovers. For example, at the end of the calendar year, inventories of cereals are near their maximum in the northern hemisphere, indeed they are just a few months into the season, after being harvested between August and November. To make a sensible assessment of exportable supplies for the year ahead, it is necessary to take the crop year or the marketing year as the base. Not all food products share the problem of seasonal production peaks. For these commodities the calendar year can be a suitable marketing year. Total stocks are usually taken to be those in the hands of farmers, food industries and the government. Household stocks are usually excluded.

Aggregation over countries presents some technical difficulties, as crop seasons vary. However, over the years, the international agencies making estimates have developed ways of aggregating that are not problematic. For some countries, they aggregate all their food grains on the same marketing year even though the individual crops are on different bases (e.g. India). The important point for international assessments is that these data can be re-aggregated on a more internationally comparable basis. The basic problem with data on carryovers of grains and oilseeds is that they are often not reported and, frequently, not collected. This applies widely to developing countries but to others too. It is difficult to have complete confidence in world carryover figures in this state of affairs. Countries that do not report need to be reminded of the importance of these data in making world food market and also food security assessments. Estimates are being made and will continue to be made by analysts in FAO, International Grain Council (IGC), USDA and so on. These estimates usually have to make at least rough calculations for an initial year, probably when consumption is at a low and it can be fairly assumed that stocks too had been run down to a minimum level. From this point, annual series of net changes in stocks (production + imports-exports-consumption) can be added and subtracted to arrive at stock level estimates for all later years. Care must be taken so that the sum of cumulative net changes in stocks does not become negative at any point, as was the case with estimates of China's stocks made by international agencies in the 1990s, which prompted a re-examination of the underlying data series of flows.

It is helpful, of course, if other sporadic stock estimates, partial or full, are available for particular years in order to refine the series. These estimates should be discussed with the countries concerned and with international experts called to special study group meetings on these questions. Private traders may also be approached to elicit alternative viewpoints on what is arguably the most difficult part of world food market monitoring.

Doubts over Consumption Forecasts

Consumption has traditionally been considered the supply/demand element that was the most stable over time; after all, stock changes and trade flows are essentially there to offset the effect of production variations on consumption. Yet, significant (and often little reported) changes do occur in consumption or consumption policies (including policies related to biofuels). In fact, at the beginning of world food crisis in the 1970s, it was the decision made by the former USSR to maintain consumption levels that caused the unexpected surge in grain imports. By stretching the export capacity of the United States of America, it was such imports that helped set off the grain price rise in 1972. Consumption trends and policies probably are one of the factors that need monitoring, as their impact can have some startling effects, especially when there are a number of closely competing commodities, as in the feed-grains/oilcakes complex (or in the vegetable oil complex) where substitution among products plays an important role. A little-regarded trade concession by the then European Economic Community (EEC) let in a flood of cassava imports to its feed market, pushing out correspondingly large volumes of feed-grains in just a few years in the early 1970s. More recently, the high petroleum price combined with government policies to foster biofuels (in the United States of America and to some extent in the European Union and, increasingly, also other countries) led to a huge diversion of grain and various oil-crops away from traditional outlets and into making fuel. Another little-researched area is the widespread use of vegetable oils as *oleo-chemicals* (other than for biofuel) where technical progress, changing consumer habits, product substitution and other factors cause the market to be very dynamic.

Monitoring of consumption, it is suggested, could focus on the following four areas. First, there is a need to undertake demand studies for the fast growing major developing country markets, as the sheer speed of economic growth means that consumption patterns may move fairly quickly away from consuming basic grains to diets that are richer in protein and other highly income elastic products. Demand studies may also be needed in the oilseeds sector on changing consumer preferences on the presence of Genetically Modified Organisms (GMOs), products produced using environmentally and socially sustainable practices (palm oil) and product health attributes (e.g. saturated fats and trans fats). These studies can be one-off studies to identify the trends to watch out for.

Secondly, data need to be collected more intensively on the non-food uses of grains and oilseeds: not only on feed use but, importantly, also on the various sectors that use vegetable oils for non-edible and non-feed purposes. Data are simply not available for some of the oilseed end-uses in the chemical industry. Changes in the mix of ingredients both in the animal feed sector and the chemical industry can lead to important changes in demand for the raw materials.

Monitoring the end-use industries with a view toward identifying possible changes in input demand would involve undertaking visits to these industries, subscribing to trade journals and discussing with traders. Possibly, technical reports would need to be commissioned to identify changes.

Thirdly, there has been a strong growth in the use of some cereals and oil-crops for making biofuels, so that currently this use accounts for some 12 percent of world production of coarse grains and close to 10 percent of global vegetable oil production. The surge in the production of maize-based ethanol was prompted by policy measures, especially renewable biofuel blending mandates in transportation fuel, as well as higher petroleum prices. The potential for large changes in that end-use is clearly considerable; the whole industry could shrink rapidly if the policies/fuel prices were to change. This sector will need to be closely monitored and up-coming legislation followed for clues as to how this demand could evolve. Fourthly, an underappreciated problem in the utilization side of the supply/demand balances is to be found in the unreliability of seeds and waste estimates. The share of a crop used for seed or, more importantly, crop share wasted can be alarmingly high. The Post Harvest Losses Information System shows, for instance, that losses of cereals in East and Southern Africa amounted to 14-17 percent in recent years. The fear is that high figures would also be found in other countries but that information is sparse. There is, therefore, an urgent need for fresh studies for the countries and commodities concerned.

Trade Policy Changes

World trade in basic food has expanded substantially (i.e. for cereals by over 40 percent between 1990 and 2010), exceeding the growth of world production and consumption (which for cereals expanded by 24 percent and 30 percent respectively over the past two decades) but it continues to vary over time, mainly because of production shocks and changes in consumption. The volume of trade is also influenced by changes in trade policies - witness the decision taken by a number of countries both in 2008 and 2010 to ban exports, which had notable impacts on market sentiment in those years.

Over the past two decades trade policies have increasingly been geared toward market opening by importers and restraint on export subsidies by exporters. Despite the major efforts made in trade liberalization in the period leading up to (as well as in consequence of) the Uruguay Round, by and large, the change in trade policies has been controlled and market opening usually gradual, at least as far as the major trading countries are concerned. For many of the smaller developing country importers, market opening was often more dramatic and food imports surged.

Although this opening helped in increasing trade, it did not cause upsets in the world market. The main causes of disturbances to world food markets in recent years have been production shocks that affect import demand or export supply, which are then reflected in trade policy adjustments. For example, there was a sharp fall in output in the Russian Federation in 2010 that preceded the export ban; the production drop would have caused, in any case, some fall in exports so the net effect of the policy change is less than the headline effect. Still, trade policy, because it acts directly on the world market, often has a psychological effect on markets that needs to be kept in mind. On the other hand, sharp changes in tariffs or export taxes can and do have real and substantial market effects, as can non-tariff barriers and changes in industry standards (e.g. sustainable certified palm oil, labelling for trans fat content). Thus trade policies need to be monitored, including ongoing negotiations under regional and international agreements

Monitoring Price Developments

Key to any food market monitoring systems is, of course, prices. But it is necessary to be clear about the type of prices involved. The most immediate concern is the current price paid by food industries and eventually consumers and received by traders and farmers. This current price can be measured by wholesale, retail, producer, import or export prices. Essentially all countries have such data, although not necessarily of the same quality in terms of coverage, frequency and representativeness. Looking for indications of prices in a few months or a year ahead there are futures markets in a number of countries where quotations are available for prices at specific dates and for specific qualities (the Chicago Mercantile Exchange in the United States of America, Euronext LIFFE, as well as those in Argentina, Brazil, China, India, Japan and Malaysia, among others). Price reports, however, are not to be found everywhere; in practice there are relatively few that are open to traders from other countries and which serve as benchmarks for world trade.

Monitoring prices, be they current or futures, is more complicated than meets the eye. There are many varieties and grades of all grains; for rice there are prices for paddy, un-milled rice, polished rice, parboiled rice, graded by percentage of broken rice, long, medium and round grains, aromatic or glutinous varieties, and so on. As far as international prices are concerned, the analyst has to select the most representative and either report these types or prepare price indices, as is now widely done. Using primarily export or import price quotes from specific ports deemed to be representative for world markets, FAO has world price indices for food, rice, oilseeds, oils and oilcakes, dairy and meat products, which are published on monthly basis. Other agencies also construct food/agricultural-related price indices not only because of the heterogeneity of most products but also because prices are often not quoted at certain times of the year or, in particular, when supplies are short. But, in this area, it appears that price monitoring, as currently undertaken, is adequate. The same cannot be said for national domestic price series, where in spite of recent efforts by FAO, improved coverage could be important, especially in the major trading countries. However, because price data are not always of the desired quality, some special efforts may be necessary to improve the flow by engaging local consultants and strengthening local capacity.

Futures prices are structured to refer to a particular date ahead; the length of time ahead that a particular contract refers to gradually shrinks with every day that passes until the contract period closes. The standard view is that the futures price converges on the spot price even though, for technical reasons, the two prices are not equal (the basis). If the classical view is correct, the futures contract typically has to be priced somewhat below the price that the market is expecting (so called normal backwardation), so that the investor in futures contracts can make a gain for the risk being taken. For these two reasons the relation between the futures price and spot prices in the future is not one-to-one. The situation is more complicated when there are next to no stocks in the market in the period when the futures contract is open. In these circumstances, the link between current spot prices and futures prices breaks down and arbitrage over time is ruled out. In other words, while futures prices are a useful pointer to the prices in a few months time, they have to be used carefully.

In recent years, concerns have grown about the influx of investment in the big internationally-orientated futures exchanges by non-commercial interests like banks and hedge funds. The importance of this phenomenon can be gauged by comparing two situations, one without non-commercial interests and another when new buyers for futures enter the market. In the first case, the sellers in the market are basically farmers and the buyers are basically food and feed industries (ignoring foreign trade). Farmers sell forward their future output at a price that they can accept and the industry receives a price at which they can do business. Both "lock-in" the price.

The stockists undertake time arbitrage so that the difference between spot and futures price is close to the cost of storage plus a "normal" profit margin. When there is an influx of investment from outside the sector, the demand for futures contracts rises and, through time, arbitrage may raise the spot price. With all the caveats mentioned above, it is clear that an influx of money from outside the commercial sector will raise both spot and futures prices; should such investments leave the market both spot and futures prices will fall. This statistic - the net long position of noncommercial operators - is a useful indicator of market sentiment and should be monitored. Some information and analysis in this regard is included in FAO's Food Outlook reports twice a year along with regular assessment of food import bills and implied volatility, but more frequent and detailed analysis are required in order to enhance transparency and market information. Moreover, other indicators should be developed and a special study should be commissioned to develop such indicators.

The Way Forward

Improved monitoring must be disseminated if it is to play its role in enhancing market transparency. There is a need for both timeliness and frequency of the outputs. The case can be made that the FAO should issue regular short updating documents to Food Outlook, as it did in earlier years. Regular publication of the supply/demand situation in tabular form, perhaps accompanying price updates and selected number of market indicators, may also be helpful. The important guide for an FAO publication is that the outputs are seen as dependable and independent of special interests. To be timely, however, some risks must be taken and judgements on complex unfolding situations may occasionally err. There is no way of completely avoiding errors or wrong judgments, but an annual, short review of forecasts analysing the performance should also be made available to readers. In addition, we suggest that the monitoring reports of policy developments indicated above be released not only because they are useful on their own, but because they can help other analysts understand the basis of FAO forecasts.

One question that has been left aside is the commodity coverage of the enhanced monitoring. Cereals and oilseeds, oils and oilcakes are discussed above, but it would be desirable to extend this monitoring to include the complex but important group of livestock products in view of their significance to world food trade and food security everywhere. Markets for these products are large and have received a boost from trade liberalization. Livestock products are, however, complicated because they are so heterogeneous, even more so than oilseeds, oils and oilcakes. Price data are often poor and, hence, the use of indices is virtually obligatory. In addition, there is the difference between systems of intensive livestock feeding and feeding on pasture. Monitoring pasture conditions is a weak point in this area and further work is needed.

References

FAO (2004). Critical review of China's cereal supply and demand and implications for world markets, CCP:GRRI/04/2, Rome.

Financial Times (2010). Food supply woes fatten Cargill earnings, 13 January.

Singer, H. W. (1997). A global view of food security. *Agriculture + Rural Development*, 4: 3–6. Technical Center for Agricultural and Rural Development (CTA).

von Braun, et.al (2004). *Agriculture, food security, nutrition and the Millennium Development Goals* (Annual Report Essay) Washington, D.C.: International Food Policy Research Institute (IFPRI)

Working, H. (1953). Futures trading and hedging, *American Economic Review*, XLIII: 318–20.

Bibliography

Abbott, J. (1987), *Marketing Enterprise For The Developing World*, Cambridge University Press, pp. 181–187.

Acharya, S.S.. Agricultural Marketing in India: Some Facts and Emerging Issues, *Indian Journal of Agricultural Economics*, Conference no. 53(3): July – September.

Agcaoili, M. & Rosegrant, M.W. (1992). *Regional trends and projections for food supply demand and trade in Asia*. Washington DC, IFPRI.

Al Hassan, R.M. (1989) *Economic evaluation of storage of grains in the Northern Region of Ghana in Effective post-harvest management of cereals and legumes in Ghana*. University of Ghana, Jan 1989.

Araullo E.V., De Padua D.B., & Graham M. (1976). R*ice postharvest technology*. IDRC, Ottawa, Canada. 394pp.

Briggle, L.W. & Curtis, B.C. (1987). Wheat worlwide. *In* E.G. Heyne, ed. *Wheat and wheat improvement*, 2nd ed. Agronomy 13: 1-32.

Chand Ramesh. *Trade Liberalisation, WTO and Indian Agriculture: Experience and Prospects*, Mittal Publications, New Delhi.

Conway, J.A., et.al. (1992) A study in the determination of quality/value relationships in rice. *NRI Bulletin* 55: Chatham, UK: Natural Resources Institute. 45pp.

Dinham, B. and Hines, C. (1983), *Agribusiness In Africa*, Earth Sources Research Ltd., London, pp. 115–119.

FAO. (2002). *World agriculture: towards 2015/30, an FAO study*. Rome.

Flinn, J.C. & De Datta, S.K. (1984). Trends in irrigated rice yields under intensive cropping at Philippine research stations. *Field Crops Res.*, 9: 1-15.

Foster, G.H. (1982) *Drying Cereal Grains. In Storage of Cereal Grains and Their Products*. C M Christensen, Ed. St Paul: American Association of Cereal Chemists Inc. 79116.

Gaedeke, R.M. and Tootelian, D.H. (1983) Marketing: *Principles And Applications*. West Publishing Company, Minnesota, p.11.

Giga D.P. and Katerere Y. (1986) *Rural grain storage in Zimbabwe: Problems, losses and prevention.* ENDA Zimbabwe and University of Zimbabwe, Harare. 96 pp.

Greeley, M. (1987) *Postharvest Losses, Technology and Employment. The Case of Rice in Bangladesh.* Boulder, Coloado: Westview Press, xviii, 345 pp.

Hanson, H., Borlaug, N.E. & Anderson, R.G. (1982). *Wheat in the third world.* Boulder, CO, USA, Westview Press.

Jewers, K., qt.al. (1989) Methodological developments in the sampling of foods and feeds for mycotoxin analysis. *Journal of Applied Bacteriology Symposium Supplement:* 105S-116S.

Jha Shikha and P.V. Srinivasan. Food Security through Price stabilisation: Buffer Stocks vs Variable Levies, *Economic and Political Weekly*, 34(46 &47): 3299-3304, November 20.

Johl, S. S.. Agriculture Sector and New Economic Policy, *Indian Journal of Agricultural Economics*, Conference no. 50(3): July -September.

Kenny, G. (2001). *Nutrient impact assessment of rice in major rice consuming countries.* FAO-ESNA Consultancy Report.

Kohli, M.M. & McMahon, M.A. (1988). A perspective of research needs for non-irrigated tropical countries. *In* A.R. Klatt, ed. *Wheat production constraints in tropical environments*. Mexico, DF, CIMMYT.

Kohls, R.L. and Uhl, J.N. (1990) *Marketing Of Agricultural Products, 6th edition, New York,* Macmillan Publishing Company pp. 18–21.

Kumar Praduman. Food Demand and Supply Projections for India, *Agricultural Economics Policy Paper* 98-01, Indian Agricultural Research Institute, New Delhi.

Leonard, W.H. & Martin, J.H. (1963). *Cereal crops*. New York, NY, USA, MacMillan Publishing.

Orth, R.A. & Shellenberger, J.A. (1988). Origin, production, and utilisation of wheat. *In* Y. Pomeranz, ed. *Wheat chemistry and technology*, vol. 3. St Paul, MN, USA, American Association of Cereal Chemists.

Peterson, R.F. (1965). *Wheat botany, cultivation and utilization*. New York, NY, USA, Interscience Publishers.

Phillips R. (1981) *Viable Grain Storage for Small Farmers in Developing Countries.* Manhattan, Kansas: Food and Feed Grain Institute, 43 pp.

Piggot, R. and D.S. Sidhu. *Efficiency of Foodgrain Marketing in India*, Background Working Paper, World Bank, Mimeo.

Pulver, E. & Nguyen, V.N. (1998). Sustainable rice production issues for the third millennium. In *Proceedings of the Nineteenth Session of the International Rice Commission*, Cairo, Egypt, p. 32-43.

Rao C.H. Hanumantha. Declining Demand for Foodgrains in Rural India: Causes and Implications, *Economic and Political Weekly*, Vol. 35 No. 4, Jan 22.

Scarborough, V. and Kydd, J. (1992), *Economic Analysis of Agricultural Markets:* A Manual. Chatham, UK, Natural Resource Institute, p. 140.

Selassie, H.G. (1995), *International Joint Venture Formation In The Agribusiness Sector*, Avebury, Aldershot.

Singer, H. W. (1997). A global view of food security. *Agriculture + Rural Development*, 4: 3–6. Technical Center for Agricultural and Rural Development (CTA).

Singh G., Bhardwaj K.C. (1985). *Directory of agricultural Machinery and Manufacturers.* Central Institute of Agricultural Engineering, Nabi Bagh, Berasia Road, Bhopal, India.

Smith, L. (1992), *Agricultural Sector Performance and Policy And the National Economy.* Network And Centre For Agricultural Marketing Training In Eastern And Southern Africa, p. 74–80.

Spooner, N. and Smith, L. (1989), *Structural Adjustment Policy Sequencing in Sub-Saharan Africa,* Rome: Economic and Social Policy Division, Food and Agriculture Organization.

Srinivasan P.V. and Shikha Jha. Food Security through Price stabilisation: Buffer Stocks vs Variable Levies, *Economic and Political Weekly*, 34(46 &47): 3299-3304, November 20.

Swaminathan Madhura. Understanding the Costs of Food Corporation of India, *Economic and Political Weekly*, 34(52): A121-AA132, December 25.

Tuong, T.P. & Bouman, B.A.M. (2002). Rice production in water-scarce environments. In *Proceedings of International Conference on Irrigation Water Management.* USDA. 2002. ERS, Database.

von Braun, et.al (2004). *Agriculture, food security, nutrition and the Millennium Development Goals* (Annual Report Essay) Washington, D.C.: International Food Policy Research Institute (IFPRI)

World Bank. *India Foodgrain Marketing Policies: Reforming to Meet Food Security Needs,* vol.I and II, Report no. 18329-IN, April 17.

Working, H. (1953). Futures trading and hedging, *American Economic Review*, XLIII: 318–20.

Yuan, L.P. & Fu. (1995). *Technology of hybrid rice production*, p. 84. Rome, FAO.

Index